Bibliography
and Footnotes

Bibliography and Footnotes

A Style Manual for Students and Writers

By PEYTON HURT

THIRD EDITION
REVISED AND ENLARGED BY
MARY L. HURT RICHMOND

UNIVERSITY OF CALIFORNIA PRESS
BERKELEY, LOS ANGELES, LONDON

University of California Press
Berkeley and Los Angeles, California
University of California Press, Ltd.
London, England

Third Edition, 1968
Copyright © 1949, 1968, by
The Regents of the University of California
Second Printing, 1973
ISBN: 0-520-00589-9
Library of Congress Catalog Card Number: 67-26633

TO

SYDNEY B. MITCHELL

IN APPRECIATION OF HIS
NEVER-FAILING GOOD NATURE
WISE COUNSEL AND
STIMULATING PROFESSIONAL LEADERSHIP

PREFACE TO THE THIRD EDITION

In the course of time since *Bibliography and Footnotes* was first printed, many students, research workers, and librarians concerned with advising writers have commended the work as a concise and practical guide. Many, too, have requested "more illustrations" and "much more about government documents." The present revision offers additional information, with as little sacrifice of brevity as possible.

The fundamentals of bibliographical and footnote references to books and articles are treated with only slight changes from what was said in the second edition. This part of the work may be used separately, without reference to the full treatment of government documents, statute and case citations, or scientific references. For the benefit of those who are primarily interested in references to documents the fundamentals of document citation and their characteristic peculiarities are described in detail in the section on United States government documents. This section may be used or taught independently of such special categories of document reference as, for example, the United Nations. The expansion of government activities and government regulation in the social and economic fields, and the attendant increase in publications reporting and describing them, have made government documents one of our most abundant sources of printed information. Students are citing documents more than ever before, and no other class of printed materials presents so many confusing problems. The document sections in this manual are sufficiently detailed, and include a sufficient number of illustrations, to enable the student to apply the principles even when no special instruction has been given in class work.

Instructions for typing the final draft of a manuscript, and specimen thesis pages, have been added to meet the needs expressed by graduate students. It was not considered practical, however, to increase the size of this book further by inclusion of material on research methods; several dependable manuals are already available on general methods of assembling data, and others treat the special subject fields. Similarly, no attempt has been made to offer detailed rules of editorial practice, correct usage, proofreading, and the like, which are adequately presented in the manuals of style included in the "Selected List of References," pages 149-150. English composition and grammar, also, are considered to lie outside the scope of this work. And no attempt has been made to describe the varying usages in foreign languages, since it is assumed that if a student has a working knowledge of the principles of citation in English the application of these principles will be a matter of judgment, in whatever language the citation be made.

I am indebted to Joel F. Walters, Senior Editor, University of California Press, for his friendly interest and helpful suggestions during the preparation of this edition. Except for his sharp eye many faults might have otherwise appeared in print. Professor Wallace W. Douglas, Department of English, Northwestern University, has kindly given permission to use the title page and parts of his doctoral dissertation. The generous cooperation of Wyllis Wright, Librarian, and Juanita Terry, Reference Librarian, Williams College, has often allowed me to work conveniently in my home and also as far away as East Africa. Through the kindness of the Library staff of University College, Nairobi, Kenya, United Nations depository materials were there made available to me.

M. L. H. R.

Williamstown, Massachusetts

CONTENTS

INTRODUCTION 1
  Advantages of the forms recommended...................... 2
  The need of taking full references........................... 3

§I. BIBLIOGRAPHY

BIBLIOGRAPHY 5
REFERENCES TO BOOKS 6
  Essential Points of Information.............................. 6
    1. Author's name... 7
      Editor; compiler; translator............................ 7
      Titles of nobility....................................... 8
      Anonymous and pseudonymous works.................. 9
    2. Title 9
      Supplementary notation following title................ 10
    3. Series and number...................................... 10
    4. Edition 12
    5. Imprint—place: publisher, date......................... 12
    6. Main pagination.. 14
      Illustrations, etc...................................... 15
  Recommended Form of Reference to Books................. 15
    Corporate author... 17
    Title entry... 18
    Unpublished dissertations................................. 19
    Manuscript materials..................................... 19
REFERENCES TO ARTICLES.. 20
  Essential Points of Information............................. 20
    1. Author's name.. 20
    2. Title of the article..................................... 20
    3. Name of the periodical................................. 20
    4. Volume number.. 21
    5. Date 21
    6. Number of the issue (ordinarily omitted) 22
    7. Pagination of the article............................... 22

Recommended Form of Reference to Articles..............	22
Title entry.................................	24
Newspaper articles..........................	24
Articles in encyclopedic works...............	26
Parts of books or sets.......................	27

REFERENCES TO GOVERNMENT DOCUMENTS.................. 29

Essential Points of Information......................	29
1. Government agency as author—U.S................	29
2. Personal author	30
3. Title	31
4. Importance of series and number...............	31
5. Edition	31
6. Imprint	32
7. Serial number	32
Recommended Form of Reference to Documents...........	33
1. U. S. Congressional publications................	33
2. U. S. Departmental publications................	35
Alternate form of document reference..........	37
Periodicals published by the government.......	37
3. *Congressional Record*..........................	37
4. U. S. laws, statutes, etc.......................	38
5. U. S. Supreme Court, etc.....................	39
6. State and municipal documents................	40
State laws.................................	41
State court reports.........................	42
City charters and ordinances................	42
7. British government documents, etc..............	43
Parliamentary publications	43
Non-Parliamentary publications	46
Parliamentary Debates......................	47
English statutes...........................	48
English court reports.......................	48
8. Other foreign government documents...........	49
9. Publications of international organizations......	50

League of Nations..	50
International Labor Organization.......................	53
Permanent Court of International Justice..............	55
United Nations Conference on International Organization	56
United Nations Preparatory Commission...............	57
United Nations	58
United Nations Sales Publications......................	64
United Nations Secretariat................................	70
International Court of Justice...........................	73
Specialized agencies.......................................	75
Regional organizations....................................	77

§II. FOOTNOTES

FOOTNOTES...	81
FULL FORM OF FOOTNOTE CITATION (For First Reference)........	84
1. Books	85
2. Articles	86
3. Newspaper articles ....;...........................	87
4. Parts of books—encyclopedia articles...................	88
5. Indirect quotations	88
6. Dissertations, manuscripts, and interviews.............	89
7. U.S.—national, state, and municipal documents........	90
8. British government documents.........................	91
9. International organizations	92
League of Nations.......................................	92
International Labor Organization.......................	94
Permanent Court of International Justice..............	94
United Nations Conference on International Organization	94
United Nations Preparatory Commission...............	95
United Nations...	95
10. *Congressional Record* and *Parliamentary Debates*........	99
11. Laws, statutes, etc.......................................	100
U. S. federal laws, etc...................................	100
State laws, etc...	101
English statutes	102

12. Court decisions	103
U. S. Supreme Court reports	103
State court reports	104
English court reports	104
SHORTENED FOOTNOTE CITATIONS (For Later Reference)	105
Use of *ibid.; op. cit.; loc. cit.*	106
Abbreviations, symbols, etc.	109
Capitalization and italics	111
Classical and literary references	111

§ III. SCIENTIFIC AND TECHNICAL REFERENCES

SCIENTIFIC AND TECHNICAL REFERENCES	113
Alternatives to Footnote Citation	114

§ IV. TYPING THE MANUSCRIPT

TYPING THE MANUSCRIPT	119
Number of copies	120
Materials: paper, etc.	121
Margins	121
Typing instructions for dissertations	122
Typing manuscripts for publication	123
Paging	123
Title page	124
Preface	125
Table of contents	125
Illustrations, etc.	126
Introduction	127
Headings	127
Footnotes	127
Bibliography	130
Appendixes	130
Index	130
SPECIMEN PAGES	134
SELECTED LIST OF REFERENCES	149
INDEX	151

INTRODUCTION

THE SCHOLARLY investigation of a subject usually requires the intelligent and systematic collection of data to be used in support of the author's interpretations and conclusions. Inevitably, most of the data so collected will derive from printed materials, and the author is expected to acknowledge his use of these materials, not only as a courtesy, but also as an indication of his willingness to subject his sources to investigation. In making acknowledgments it becomes necessary to utilize, in bibliography and footnotes, certain techniques and conventions of citation. This manual is concerned with the forms in which the techniques and conventions are employed when proper credit is assigned by the author for the sources upon which he has relied. The basic principles which underlie the citation of data are summarized; a sufficient number and variety of examples are presented to illustrate the practical application of the principles; and the commonly acceptable variants are indicated.

The documentation of a work (i.e., citation of sources) indicates the authorities upon which the text is based. The authorities may be listed in an appended bibliography, or cited in footnotes, or both. Ideally, the bibliography in scholarly works gives the detailed identification of all works consulted, while the footnote is a specific reference apprising the reader of the exact source of a statement within the text. A clear understanding of the differences between these two basic forms is necessary to an understanding of the principles of citation.

Bibliographical references and footnote citations contain the information which permits the reader to test the origin of all important statements which are neither common knowledge nor original with the author. The items of the bibliography are general references to entire works, but the footnotes are particular references acknowledging ideas, statements, facts, or quotations taken from other authors. Footnotes may be employed to acknowledge the source of the information used, whether it be fact, theory, or opinion; to assign credit; to cite authority for controversial statements; to support arguments; to guide the reader to additional and related material; to identify quoted material; to interpret, limit, evaluate, describe, compare, or augment meaning within the text; or to refer to other pages or passages in the same text. In order that references of the one kind or the other shall be meaningful and intelligible to the reader, a clear, concise, consistent, and non-repetitive form of citation must be employed.

There is no common agreement among scholars, editors, bibliographers, or publishers concerning the exact forms to be used in making either footnote citations or bibliographical references to books, articles, or other printed works. Consequently, in the absence of any uniform style of citation, it is imperative that all the essential information for the identification and location of the materials cited be presented in some manner which does not allow of misinterpretation, and that this information be full enough to be understood by those for whose use it is intended.

The essential information which is required in references to books, articles, and documents is not difficult to assemble. However, the arrangement and sequence of the data, the punctuation, the use of italics and parentheses, the inclusion or omission of individual items, the use of abbreviations for volume, page, and number, and other details are treated variously, and reflect usages peculiar to the different subject fields. For example, in scientific works the footnote references are written commonly in a shortened form, and often eliminated by the use of various devices. In legal publications the footnote citations are given in an abbreviated style characteristic of the law. In works in the social sciences and the humanities the footnotes are usually cited in a longer form of reference which gives complete bibliographical information the first time a work is cited. In all fields of knowledge the footnote and bibliographical references reveal various forms, whether these be full or abbreviated.

Advantages of the forms recommended.—The style of citing references recommended in this manual represents standard practice, so far as any practice in current use may be said to be standard. The forms include information sufficient for the identification of the works to be cited, and have been simplified as much as clarity will allow. This information is developed in logical sequence and is nonrepetitive. Because all essential information is included in the recommended forms of bibliographical reference, these forms may easily be adapted to the prevailing practices in various fields. They may also be used with little change, either in a footnote at the bottom of the page or in a bibliography at the end of a chapter or article. However, note is made of the different forms in common use, and various methods of handling particular details are discussed.

It is recognized that, after a fair knowledge of the principles of citation and an appreciation of the significance of the individual items of a reference are acquired, the reader will probably adopt modifications which apply to his own subject or personal needs. Nevertheless, alterations should be made only with an awareness of good bibliographical practice and in conformity with it. It must always be understood that *in citation there is no one commonly accepted form;* many forms of references are permissible. But when a writer chooses one form or another, *he must be consistent within the same manuscript,* and he should include in his references at least enough particulars of information to permit identification of the works referred to. Although various methods of citation are approved by usage, the student will profit by the adoption and consistent use of one style which conforms to his personal needs and to the requirements of his subject field.

The need of taking full references.—When assembling material for a paper, thesis, or the like, a student should learn to make a complete notation of each publication used, and should cultivate the habit of so doing when using a book or article for the first time. In the daily routine of scholarship much time is lost in the relocation of printed materials because notation of the author's name, the title of the work, or the facts of its publication was erroneous or incomplete. Time is needlessly wasted, for example, from the common failure to note an author's name in full. Without the given names or initials, one must often check over many authors of the same surname in a library catalog, bibliography, or periodical index, in order to complete the notation. The details of each reference consulted by the student should be recorded with accuracy and completeness. At first this practice will require the exercise of conscious and deliberate attention to each item in the reference. Nevertheless, habits of precision are the basis of scholarship and demand cultivation.

"Does the author aim at money and profit? It will be a wonder then if he succeeds, since he will only stitch it away in great haste like a tailor on Easter-eve; for works that are done hastily are never finished with that perfection they require."[1]

---

[1] *Don Quixote*, translated by Charles Jervas, World's Classics (London: Oxford University Press, 1928), Bk. II, chap. iv, p. 36.

BIBLIOGRAPHY

BIBLIOGRAPHY is here intended to mean a list of references to publications such as may be found at the end of a chapter or at the end of a book or article—a list of the works of an author, or of the literature bearing on a particular subject. Scholarly works characteristically list completely all works consulted during the preparation of a paper or book; this list is appended, with the heading "Bibliography." The heading "Literature Cited" may be used, however, when the list includes only those references which have been cited in the footnotes. Popular works and textbooks often include only a selected list of references which may prove helpful to the reader or student; such a list is properly headed "Selected List of References."

The list of references is most conveniently arranged by author in alphabetical order, but it may be arranged chronologically by date of publication. A long bibliography may be divided into primary (source) materials and secondary materials dealing with the subject; classified according to subdivisions of the subject; or grouped, presenting separate lists of books, articles, government publications, and other categories of printed or manuscript materials. The choice of arrangement and types of subdivisions will be determined by the scope, emphasis, and subject matter of the bibliography and the amount of material to be included. Term papers and other short works will rarely require other than a straight alphabetical arrangement by author. The items of the bibliography may be accompanied by annotations which briefly describe and evaluate the work.

The items in a bibliography, either at the end of a chapter or at the end of a book or article, are sometimes numbered. Corresponding numbers are then used within the text to refer to the bibliography, and the usual footnote references are eliminated. This manual recommends that wherever possible the footnote and bibliographical references be treated separately (cf. Section III, "Scientific and Technical References").

Students should cultivate an intimate acquaintance with bibliographical tools, including the library card catalog, but discretion must be exercised in the adoption of library usage. The bibliographical styles employed in library practice are designed to meet special needs, and are likely to be more complex than is advisable or necessary in general bibliographical practice. The student is urged to develop a clear understanding of the principles underlying bibliographical citation. The application of these principles should then be tempered with judgment and common sense combined with a knowledge of the practice peculiar to his own subject field. The careless appropriation of unsuitable forms of reference, either as found in the sources consulted or on the library catalog cards, will develop harmful bibliographical habits.

REFERENCES TO BOOKS
Essential Points of Information

Of necessity, many of the remarks made in this section on books will apply also to references to articles, documents, etc. Except where the problem is a special one, the discussion in the subsequent sections will not repeat the information noted here.

Bibliographical information should be taken from the title page of the book. This information should never be taken from the binding, since cover design is determined by attractiveness and available space, and not by completeness or accuracy. If necessary, it is permissible to supplement the information on the title page, but the supplementary data should be enclosed in square brackets to indicate that the information was not found on the title page. The proper use of square brackets is discussed below under each item of the citation. This refinement, i.e., square brackets, is not obligatory in term papers and general works, but should be used in dissertations and scholarly manuscripts intended for publication.

The library card catalog will ordinarily provide information which is lacking on a book's title page. Another valuable source is *A Catalog of Books Represented by Library of Congress Printed Cards*[1] and its successor *The National Union Catalog.*[2]

---
[1] (Washington: Library of Congress, 1942–55). The title of the work is omitted in this footnote because it is given in the text; see also p. 83.

[2] (Washington: Library of Congress, 1956–   ).

Bibliographical terms found on the title pages of foreign books may present difficulties. English equivalents of library and bibliographical terms in foreign languages will be found in several of the books listed on pages 149-150 below.[3] Many of the works listed present detailed rules for capitalization, syllabification, etc.[4]

The following are the essential points of information which should be noted in full bibliographical references to books.

1. AUTHOR'S NAME

The author's name is written in inverted order, i.e., surname first, so that references in the bibliography may be arranged alphabetically. The given names should follow a comma after the surname (Hopkins, Gerard Manley). The full form of the first given name helps to identify the author and enables the reader to locate the work in a library card catalog more conveniently. Initials may be given if the author has more than one given name (Keynes, J. M.), but not if he has only one given name (Welles, Sumner; *not* Welles, S.); nevertheless, the full form (given names spelled out) is preferable, especially if the surname is a common one (e.g., Smith, Justin Harvey; *not* Smith, J., or Smith, J. H., which might stand for names far removed, in a card catalog, from the correct one). The author's name is followed by a period (unless it stands on a line by itself). When a book is written by two or three authors (i.e., joint authors), the name of the author listed first on the title page is given in inverted order, and the others follow in regular order (Winterich, John T., and David A. Randall, *or* Hoffman, Frederick J., Charles Allen, Carolyn F. Ulrich). When there are several authors, the first may be given with "and others" (or: *et al.*) to indicate collaborators (Platner, John W., and others). Professional titles and degrees (e.g., Captain, the Rev., M.D., Ph.D.) are not included in references.

---

[3] *United States Government Printing Office Style Manual* (rev. ed.; Washington: G.P.O., 1967), pp. 387–492, gives selected bibliographical terms which should be adequate for a student's needs.

[4] *Ibid.*, pp. 21–56.

Editor; compiler; translator.—When a book is to be listed under the name of the editor, compiler, or translator, this should be shown by the use of proper abbreviations (Hudson, Manly O., ed.; Kane, Joseph N., comp.). Most books requiring such an entry (i.e., first item in the reference) will be fairly obvious and will not present any choice—collections of the works of several authors, collections of treaties or statutes, commentaries, books of quotations, etc., are almost invariably entered by editor; but where the editorial comment is generally considered as significant as the text itself (e.g., Birkbeck Hill's exhaustive editorial comment in the Oxford edition of Boswell's *Life of Johnson*), or where the translation carries its own intrinsic value (e.g., Pope's translation of the *Iliad* or the *Odyssey*), the choice of entry depends upon the nature of the bibliography and the viewpoint which the author wishes to emphasize. *The Letters of William and Dorothy Wordsworth*, edited by Ernest De Selincourt,[5] may be listed under Wordsworth or under De Selincourt, depending upon which aspect the maker of the citation wishes to accentuate. Laurence Binyon's translation of Dante's *Purgatorio*[6] may be listed with equal validity under either Binyon or Dante; in a book on modern poets it would appear to better advantage under Binyon's name, but in a book on Italian poetry it could only be listed properly under Dante. Rules cannot be prescribed to cover all special circumstances; the choice of entry must be the author's own and should reflect his approach to the subject. Consistency within the same manuscript is indispensable.

Titles of nobility.—Similar problems of choice are encountered when an author holds a title of nobility. It is best to choose the form of the name by which the author is better known, whether this be the family name or the title. John Buchan (Lord Tweedsmuir), Benjamin Disraeli (Earl of Beaconsfield), and Horace Walpole (Earl of Orford) are known by their family names; Lord Halifax (Edward Frederick Lindley Wood), Count Rumford (Benjamin Thompson), and Lord Beaverbrook (William Maxwell

---

[5] (Oxford: Clarendon Press, 1937–39), 5 vols.
[6] *Collected Poems* (New York: Macmillan, 1931), II, 187–211.

Aitkin) are seldom referred to except by their titles. Detailed rules of usage are not practical; the choice should be regulated by common sense and judgment based upon some acquaintance with printed references to the person. The fullest forms of the names will be found in the library card catalog; this information will be helpful, but strict adherence to library usage is not advised. The forms of the names used in the standard encyclopedias and biographical dictionaries will prove to be safer guides.

Anonymous and pseudonymous works.—If the author of an anonymously published book can be identified, the book should be listed under the author's name enclosed in square brackets, to indicate that the name was not found on the title page.[7] When the name of the author cannot be determined, the book should be listed by title (see page 18), although some prefer to list it under the word "Anonymous" in author position. Pseudonymous works may be entered under the pseudonym, with the real name of the author following and enclosed in square brackets (e.g., Nadar, *pseud.* [i.e., Félix Tournachon]; *or:* Twain, Mark [pseud. of Samuel L. Clemens]), or entry may be made under the author's real name with the pseudonym following (e.g., Brontë, Charlotte [*pseud.* Currer Bell]; *or:* Beyle, Henri [*pseud.* Stendhal]). In general, the better-known form of the name should be preferred. In a lengthy bibliography, a cross reference from one form of the name to the other would be helpful to the reader. Where the pseudonym or a shortened form of the name has become so well known that reference to the real name would be confusing, omit any explanation; e.g., the real name of Norman Angell is Ralph Norman Angell Lane (see illustration, p. 16), and Robert Laurence Binyon is the full form of Binyon's name (see p. 8), but the shorter and more familiar forms are the preferable ones.

2. TITLE

Titles of books are written in italics, that is, underlined in type-

---

[7] If the authorship of an anonymously published book is indicated by plausible evidence, the following kind of indication is sometimes made: Dunlap, William [supposed author], *Rinaldo Rinaldini; or the Great Banditti*, a tragedy in five acts, by an American and a Citizen of New York (New York: Printed for the Author, 1810).

script or manuscript; the first word and the main words of the title are capitalized. Do not capitalize articles, prepositions of fewer than six letters, or conjunctions unless they appear as the first word of the title. The title should be taken from the title page of the book. Punctuation marks in the printed title should be preserved. A modified or shortened title usually appears on the cover and therefore is likely to be incorrect if used in a bibliographical reference. There are some few publications (usually of the pamphlet type, e.g., Foreign Policy Association, *Headline Series*) which carry the full title on the cover of the publication and only a modified title on the title page. Publications of this kind will be rare and fairly obvious, but judgment should be used in the choice of the title copied. The safest general practice is to take the title from the title page. Subtitles are omitted unless they qualify or add materially to the meaning of the title. If the subtitle is included in the reference, the punctuation on the title page should be preserved. But if it is printed without punctuation, or differentiated only by smaller type on the title page, it may be set off in the reference by a colon, semicolon, or comma, as good style and good sense may require. If the first word of the subtitle is an article, it would normally be capitalized after a colon, but not capitalized after a semicolon or comma. The title, with or without the subtitle, is followed by a period.

Supplementary notation following title.—A note immediately following the title may be used in order to give additional information, e.g., Edited by Sir William Beveridge; Translated by C. K. Scott Moncrief; Illustrated by Rockwell Kent; or Introduction by Bertrand Russell. In contrast with the title, these notes are *not* written in italics (see Kasner and Masters, p. 16). If the editor's name, or any name other than the author's, appears in author position, the information should not be duplicated in this manner. The supplementary note is followed by a period.

3. SERIES AND NUMBER

The proceedings, transactions, and reports issued by societies, the publications and studies of universities, the bulletins of government agencies, and similar series include many different titles, but the works within these series usually relate to some common field of knowledge. The series title is usually found in its most nearly complete form on the series half title (a right-hand page preceding the title page). Individual issues of a series are customarily numbered consecutively and so arranged that they may be bound together to form a set of volumes on library shelves. A reference to an individual title contained in a series of this kind should give the name of the series and the number of the particular work. Some so-called series are not significant and therefore not worth noting, as, for example, an unnumbered publisher's series for which no editor is named. In general, however, series are important and should be noted. The information is particularly significant when one is citing United States government documents, United Nations documents, etc. (see pp. 31 ff.), since the series is often the only means of identifying the item in a library card catalog.

The series and number are separated by a comma and should follow the title or supplementary note in a bibliographical reference, and for contrast *should not* be written in italics: University of California English Studies, XIV. (See Nitze and Singer, pp. 16, 17.) Many authors, however, do write the name of the series in italics (Governors' Conference, *Proceedings* for 1943; Yale Historical Publications, *Studies,* no. 14; *Frontiers of Chemistry,* Vol. IV; Boston Museum of Fine Arts, *Museum Extension Publications, Illustrative Set,* no. 4).[8] Others choose to enclose the name of the series within quotation marks (e.g., "Harvard Workshop Series," no. 5; "America Faces the Air Age," vol. 1; Dunlap Society, "Publications," n.s., no. 2).[9] Still others place the series note within parentheses.[10] Although permissible, italics, quotation marks, or parentheses are not needed to set off the name of a series in the form of reference outlined in this manual; they are not recommended, except when a general reference to the series itself is

---

[8] Homer Cary Hockett, *Introduction to Research in American History* (2d ed. with Corr. and App.; New York: Macmillan, 1950), pp. 18–19.

[9] *A Manual of Style* (11th ed.; Chicago: University of Chicago Press, [1952]), secs. 52, 79.

[10] Livia Appel, *Bibliographical Citation in the Social Sciences* (3d ed.; Madison: University of Wisconsin Press, 1949), pp. 17–18.

made within the text (p. 10 above). The position of the series note also varies; some editors and authors prefer to give the series note as the last item in the reference. The sequence recommended here places the series note immediately after the title (or note following the title) and terminates it with a period.

4. EDITION

In bibliographical references, the edition, if other than the first, should be designated in abbreviated form, using arabic numerals (e.g., 2d ed.; 6th ed.; rev. ed.).

The edition name and the name of the editor are often of primary importance in references to the classics or works by standard authors. A complete reference to a play of Shakespeare's should include a notation of the edition, e.g., New Variorum, ed. by H. H. Furness; Friendly Edition, ed. by R. J. Rolfe (cf. "Classical and Literary References," pp. 111–112). The Globe Edition is used traditionally for line references, and does not need identification in this connection. The edition designation precedes the imprint.

5. IMPRINT—PLACE: PUBLISHER, DATE

The full imprint gives place, publisher, and date of publication. This information is found ordinarily on the title page. If any part of the information is lacking on the title page but can be found elsewhere in the book itself, in the library card catalog, or in a bibliography, this supplementary information may be used, enclosed within square brackets (e.g., [priv. print.]; [copyright 1967]; [Chicago]).

The place of publication and the publisher's name may be separated by a colon or a comma. Although the colon is recommended and has been used in examples in this manual, its use is not considered obligatory. Some publishers prefer the comma.

If one publisher is given with several American cities as places where the publisher maintains offices, only the actual place of publication should be noted; thus, Boston should be noted for Houghton Mifflin, though New York may be the first place named. Two places should be included in a reference, however, if the title page shows that the publisher has both foreign and American offices (e.g., New York and London: Putnam's, 1962). When the place of publication is not well known, or if there are more than two cities of the same name, the name of the state or country should be given if identification is not clear in the rest of the reference (e.g., Portland, Me.; Cambridge, Eng.). If two publishers are given on the title page, only one need be noted in a bibliographical reference (probably the American rather than the foreign publisher); both may be given, however, in the following manner: London: Duckworth; Philadelphia: Lippincott, 1965. If no place is given on the title page, and the writer cannot find the information otherwise, the abbreviation "n.p." (no place) may be used.

Publishers' names have peculiarities of their own; Houghton Mifflin is really two names although written as one; both Harcourt, Brace and Little, Brown are punctuated by commas; McGraw-Hill and Appleton-Century-Crofts are hyphenated. The correct spelling and punctuation of publishers' names may be found in the trade catalogs, available in any good library. See, for example, the *United States Catalog*[11] and its supplement, *The Cumulative Book Index*,[12] and *Books in Print*,[13] for an exhaustive list of publishers. The annual volume of the *Bulletin* of the Public Affairs Information Service[14] (better known as P.A.I.S.) is also helpful.

To shorten the full imprint, without significant loss, it is recommended that only the key word or words in the publisher's name be given (e.g., New York: Macmillan, 1962; *not* New York: The Macmillan Co., 1962). In the student's personal notes, and even in a finished bibliography, the use of intelligible abbreviations is permissible (e.g., Wash.: Gov't Print. Off., 1961; *or* Wash.: G.P.O., 1961).

Often the name of the publisher is not included in a reference to books published more than twenty years ago, since it is likely that the books are no longer in print (i.e., available from the publishers). The name of the publisher, however, is of value when one is

---

[11] (4th ed.; New York: Wilson, 1928).
[12] (New York: Wilson, 1933–   ).
[13] (New York: Bowker, 1948–   ).
[14] (New York: Wilson, 1915–   ).

tracing an out-of-print book for purchase, and also helps to differentiate the original publication from a reprint edition. Reprint rights are often sold; consequently the publisher and date named on the title page of a reprint may vary from those named on the original. Moreover, in reprints the paging is likely to be at variance with that of the original. For purposes of identification, and to avoid possible confusion in page references, it is recommended that the full imprint be given, even for books published more than twenty years ago. The inclusion of the publisher's name may save much searching since permission for the use of direct quotations must generally be obtained from the publisher or author. Because books may be copyrighted for a period of 56 years, even those out-of-print may still be protected. (Pending legislation would extend the term of copyright to 50 years after the author's death.)

If no date appears on the title page, the copyright date may be taken from the verso of the title page, and occasionally it may be necessary to use a date found elsewhere in the book. Square brackets are used to indicate that this supplementary information was not found on the title page; e.g., [c. 1967]; [pref. 1965]; or simply [1967]. If after a reasonable search, with use of the tools suggested, no publication date has been found, the lack may be indicated by use of the abbreviation "n.d." (no date).

In works of more than one volume, it is advisable to indicate the dates of the first and the last volume, if these differ (see Angell, p. 16). In theses and term papers, shortened dates, e.g., 1957–58, are sufficient, but the use of complete dates is safer in manuscripts intended for publication, e.g., 1957–1958. Works in process of being published may be so designated by the use of a dash after the date of the first volume (e.g., 1961–   ). If only one volume of a many-volume set is used, only the date of the volume consulted, not the dates for the entire set, should be given.

6. MAIN PAGINATION

A bibliographical reference to a book should include the main pagination, which is (*a*) the total number of pages as represented by the last printed page number (685 pp.), or (*b*) the total number of volumes, if more than one (2 vols.). However, if only part of a set is used, one should list only the volumes actually consulted (see Angell, p. 85). This information is valuable in a bibliography, since it indicates the length of the work, e.g., a 25-page pamphlet, a 90-page monograph, a 450-page book, or a work in 3 volumes. A notation that a work is in two, three, or more volumes is sufficient, without an indication of the number of pages. In the Singer illustration (p. 17) the preliminary pages have been included since these form an important part of the book and exceed in length the regularly paged text. The abbreviations pp., vols., are not capitalized, and arabic numerals are used to indicate the total pagination or number of volumes. Many authors place the total number of volumes before the imprint, but inconsistently place the total pagination as the last item in the reference. Also, many bibliographies omit the main pagination in references. It is recommended that the total number of pages or volumes be included in bibliographical references as the last item in the reference (unless the illustrations as described below are included).

Illustrations, etc.—A notation indicating the presence of illustrations, maps, charts, graphs, musical scores, etc., is recommended if they are important to the subject, as in art and history, and particularly when they form a sizable part of the work cited. This item is optional, but its inclusion serves to indicate the treatment of the subject and to guide the reader directly to such materials when they are required. Often, the name of the illustrator in the supplementary note following the title (see p. 10, and Kasner, p. 16) will suffice; it is not then necessary to duplicate the data in a separate item. If this information is included in the reference, it is placed at the end, following the main pagination.

Recommended Form of Reference to Books

Bibliographical references to books should contain all the essential points of information noted above. The title is italicized, i.e., underlined in typescript or manuscript; main words of the title are capitalized (not articles, conjunctions, or prepositions of fewer than six letters); only the key words are needed in place of the publisher's full business name; indication of illustrations is optional; punctuation is exactly as indicated; spacing is in accordance with the rules of typewriting. All lines after the first line of the bibliographical reference are indented at least five spaces, for the best effect. This "hanging indention" is used in order to set off the author's name for convenience in alphabetical arrangement and ease in finding; and the prominent position thus given the author's name is often referred to as "author position." References should be written in the following form.

```
Author's Surname, Given Names or Initials. Title of Book,
in italics. Supplementary note, if necessary. Series and
number, if any and if significant. Edition, if other
than the first. Place: Publisher, Date. Main pagination,
or total number of volumes. Illustrations, etc.
```

Illustrations

```
Angell, Norman, and others. Economic Principles and Prob-
  lems. 4th ed., edited by W. E. Spahr. New York: Farrar
  and Rinehart, 1940-41. 2 vols.
Cervantes Saavedra, Miguel de. Don Quixote. Edited by Wil-
  liam Dean Howells; [translated by Charles Jervas]. New
  York: Harper, 1923. 933 pp.
Kasner, Edward, and James Newman. Mathematics and the Im-
  magination. With drawings and diagrams by Rufus Isaacs.
  New York: Simon and Schuster, 1943. 330 pp.
MacShane, Frank. Many Golden Ages: Ruins, Temples, & Mon-
  uments of the Orient. Tokyo and Rutland, Vt.: Tuttle,
  [1963]. 264 pp. 111 pls.
Masters, Dexter, and Katharine Way, eds. One World or
  None. Foreword by Niels Bohr; introduction by Arthur H.
  Compton. New York: McGraw-Hill, 1946. 79 pp.
Mencken, Henry L. The American Language: An Inquiry into
  the Development of English in the United States. 4th ed.
  corr., enl., and rewritten. New York: Knopf, 1936. 769,
  xxix pp.
---- ----  Supplement II. New York: Knopf, 1948. 890, xliii
  pp.[15]
Nitze, Paul H. U.S. Foreign Policy, 1944-45. Headline
  Series, no. 116. [New York: Foreign Policy Association],
  1956. 62 pp.
Persons, Stow. Free Religion, an American Faith. Yale His-
  torical Publications, Miscellany, no. 48. New Haven: Yale
  University Press, 1947. 168 pp.
```

---

[15] Four hyphens may be used to indicate the repetition of an author's name; in this reference, two sets of four hyphens indicate repetition of both the author's name and the title of the book. The lower-case roman numerals here indicate that the index is paged separately from the text.

Singer, Charles Joseph, and C.B. Rabin. **Prelude to Modern Science: Being a Discussion of the History, Sources, and Circumstances of Tabulae Anatomicae Sex of Vesalius.** Wellcome Historical Medical Museum, Publications, n.s., no. 1. Cambridge, Eng.: University Press, 1946. lxxxvi, 58 pp. 6 pls.

Vance, Ethel, **pseud.** [i.e., Grace Zaring Stone]. **My Son is Mortal.** London: Collins, 1961. 256 pp.

Corporate author.—In the absence of a personal author, it is often necessary to list a book in a bibliography under the name of the institution responsible for its publication. When the book is the publication of a society, government agency, institution, or other corporate body, it must be listed under the official name of the organization. (Government documents are treated separately on pp. 29 ff.) Otherwise, the form is the same as for the personal author.

Illustrations

American Medical Association. Bureau of Medical Economics. **Medical Care in the United States: Demand and Supply, 1939.** Chicago: American Medical Association, [c. 1940]. 140 pp.

    The imprint may read, Chicago: The Association, because the full name of the association appears in author position.

Chicago. University Press. **A Manual of Style.** 11th ed. Chicago: University of Chicago Press, [1952]. 522 pp.

    The imprint may read simply University of Chicago Press, omitting Chicago as place of publication, because it is obvious.

Modern Humanities Research Association. **Annual Bibliography of English Language and Literature.** Vol. XX, 1939. Edited for M.H.R.A. by Angus Macdonald and Leslie N. Broughton. Cambridge, Eng.: University Press, 1948. 292 pp.

Museum of Modern Art. **New Spanish Painting and Sculpture: Rafael Canogar** [and Others]. Exhibition. Garden City: Doubleday, [1960]. 59 pp.

Theatre Guild. **Theatre Guild Anthology.** With an introduction by the Board of Directors. New York: Random House, [c. 1936]. 961 pp.

Virginia Historical Society. **Portraiture in the Virginia Historical Society.** With notes on the subjects and artists by Alexander Weddell. Richmond: Virginia Hist. Soc., 1945. 192 pp. Pls.

The correct form of writing the name of an institution or society (the correct form of entry) in a bibliography may be difficult to determine. For examples the student should consult the library card catalog, the *Catalog of Library of Congress Printed Cards*,[16] *The National Union Catalog*,[17] the *United States Catalog*[18] and its supplement the *Cumulative Book Index*,[19] the *Union List of Serials*,[20] or bibliographies in the field.

Title entry.—Occasionally, books are written collectively and the responsibility for the publication cannot be assigned. It then becomes necessary to list the book under its title; i.e., the title assumes author position in the arrangement.

Contemporary Europe: A Study of National, International, Economic, and Cultural Trends. A Symposium by René Albrecht-Carrié and others. New York: Van Nostrand, 1941, 670 pp.

Liberty and Learning. Essays in Honor of Sir James Hight. Christchurch: Whitcombe and Tombs, 1950. 208 pp.

Although the work commonly called "Chicago Manual of Style" will be found in library card catalogs under the University of Chicago Press with the entry arranged as above (example under corporate author), it is more conveniently listed in bibliographies and referred to by the following title:

A Manual of Style. 11th ed. Chicago: University of Chicago Press, [1952]. 522 pp.

When the authorship of an anonymously published book cannot be ascertained, the book should be listed by title (see p. 9). Dictionaries and encyclopedias should also be listed under title unless the work is generally referred to by the editor's name.

Unpublished dissertations.—References to unpublished theses or dissertations follow the form used for books, except that the regular imprint is replaced by a statement that the work is an unpublished thesis (M.A., M.S., Ph.D.); the name of the institution granting the degree; and the date of the acceptance of the thesis. (See specimen title pages, pp. 135, 140.) Some publishers, preferring to restrict their use of italics to printed works, leave the titles

---

[16] (Washington: 1947–55).
[17] (Washington: 1956–   ).
[18] (4th ed.; New York: Wilson, 1928).
[19] (New York: Wilson, 1933–   ).
[20] Edna B. Titus, ed. (3d ed.; New York: Wilson, 1965), 5 vols.

of theses and dissertations unitalicized, and usually enclose them in quotation marks. The pages of a typewritten manuscript are usually designated as numbered leaves.

Illustrations

Hamilton, Daniel H. <u>DDT</u> <u>Tolerance</u> <u>Levels</u> <u>in</u> <u>Small</u> <u>Animals.</u>
 Unpublished M. S. thesis. Williams College, 1964. 40 numb. leaves., charts.

Shumaker, Charles W. <u>English</u> <u>Autobiography:</u> <u>Its</u> <u>Materials,</u> <u>Structure,</u> <u>and</u> <u>Technique.</u> Unpublished Ph.D. dissertation. University of California, 1943. 273 numb. leaves.

Manuscript materials.—Reference to manuscript materials should include enough particulars of information to identify the manuscript and indicate its location.[21] Ordinarily the location (place of deposit), the name of the collection, and the number of the manuscript will be sufficient identification, if the manuscript is contained in a well-organized collection. When the authorship of a manuscript is known, it is listed in the bibliography under the name of the author. Individual manuscripts frequently lack a title, but the descriptive title assigned by a cataloger may be included in a reference and is enclosed in quotation marks. A title invented by the author citing the manuscript is not enclosed in quotation marks and should be accompanied by an explanatory note. If the manuscript carries a date, it should be included, but if the date is assigned on the basis of evidence, it should be enclosed in square brackets. When the original manuscript is not used, the citation should specify the type of reproduction, i.e., facsimile, photostatic copy, or typewritten or handwritten copy. Manuscripts are unique and often cannot be adapted to any uniform system of citation. The reference may require descriptive or explanatory notes, or other information which would be unnecessary if one were citing printed materials. The examples given below illustrate bibliographical references to manuscripts which are part of well-organized collections. The sequence of items in the reference may be varied to accord with the text discussion. Manuscripts privately held by individuals would require further information. References to unpublished letters are usually made in the footnotes. If the

---

[21] Appel, *op. cit.*, pp. 26–27.

letters are part of a named collection and can be so identified, a general reference may be given in the bibliography (see the fourth illustration below; cf. pp. 89–90).

Illustrations

"Amadís de Gaula." *Lansdowne MSS 766.* British Museum. London, Eng.
   May be cited *MS Lansdowne* 766.

John Paul Jones Manuscripts, 587. Peter Force Collection. Library of Congress. Washington, D.C.

Robinson, Edwin Arlington. "The Man Against the Sky." *Snyder Collection.* Williams College. Williamstown, Massachusetts. 8 MS leaves, autographed.

The Montana Papers. 1919-1920, nos. 155-77. William Andrews Clark Memorial Library. Los Angeles, California.

REFERENCES TO ARTICLES
Essential Points of Information

Full bibliographical references to articles in magazines, learned societies' journals, and other periodical publications require forms somewhat different from those outlined for books. The following are the essential points of information to be noted in full bibliographical reference to articles.

1. AUTHOR'S NAME

Written in inverted order; the same rules apply as described for bibliographical references to books, pages 7–9 above.

2. TITLE OF THE ARTICLE

Titles of articles are written within quotation marks. This usage clearly distinguishes articles from books, the titles of which are written in italics (underlined in typescript or manuscript). Quotation marks indicate at a glance that the article is included in a larger work (the periodical or journal) which is the item to be requested in a library. Main words in the title are capitalized, and the title is followed by a comma placed within the quotation marks.

3. NAME OF THE PERIODICAL

The name of the periodical is written in italics because it is the complete work which contains the article to which reference is made. In the bibliography the name of the periodical should be written in full (cf. pp. 86–87). Where changes in the name of a periodical occur, the name carried at the time the article appeared is the one to be used.

4. VOLUME NUMBER

Issues of a periodical are usually numbered, in order that several of them may be bound together to make a volume; the back file in a library will constitute a bound set of numbered volumes. The volume number is commonly written in (capital) roman numerals (XXXVIII), although some periodicals use arabic numerals (38). Traditionally, bibliographical references are written with the volume number in roman numerals, but some publishers now prefer arabic numerals. It is a safe practice to use the form of the number found on the periodical itself, roman or arabic as it happens to be. In order to be consistent in the final draft of a manuscript, it is desirable to unify the volume numbers, using only one form. If changes are made, care must be exercised when numbers are switched from one form to the other.

Periodicals sometimes begin, at intervals, a new sequence of volume numbers. Each sequence is then numbered as a series, or labeled simply "new series" (abbreviated n.s.). Identification of such series should accompany the volume number (see Horne and Nosworthy, p. 23).

5. DATE

The exact date of the issue which contains the article should be given: the month and the year for monthly or quarterly works (July, 1965); and the month, day, and year for periodicals issued more frequently (July 15, 1965). The date indicates the timeliness of an article and, with the volume number, doubly identifies the volume of the periodical in which the article is published; but many scholars give only the year in a reference (1965). The practice of omitting the month and day is not recommended, especially for the author's initial references and his personal notes. It is easy to strike out material to shorten references in the final preparation of a manuscript, but it may be difficult to check back to find and insert the missing items.

When a periodical is published quarterly, with the designation Spring, Summer, Autumn, and Winter (or First Quarter, etc.), it is advisable to include this information in the reference.

Some periodicals do not number the pages continuously throughout a volume composed of several issues, but treat each issue separately, beginning with page 1, as for example the *Saturday Evening Post, Time,* etc. Care should be taken to give the exact date of the issue in a reference to a periodical of this type, since the page numbers will be duplicated in several different issues within the volume cited.

6. NUMBER OF THE ISSUE (ORDINARILY OMITTED)

The issues that make up a volume of a periodical are usually numbered consecutively. Ordinarily, the number of the issue should not be included in a reference. The date and pagination may be relied upon for identification of the issue and location of the article when necessary (see *Business Week* example, p. 24).

7. PAGINATION OF THE ARTICLE

Since nearly all periodicals are paged continuously throughout each volume, the volume number and inclusive paging will often suffice to identify and locate articles. The figures in page references may be written in the shortest form consistent with clarity of meaning, for example, 113–24, 290–311 (*not* 290–11), 1141–8, 1190–1235 (*not* 1190–35 or 1190–235). Although the shortened form of page references is permissible, in a manuscript intended for publication the safest practice requires that page references repeat the whole number, up to four digits, e.g., 76–79, 231–237, 1562–1565. If an article is not on consecutive pages, the exact pagination should be shown, e.g., 135–39, 185–206, 244. (See *Notes and Queries* example, p. 24.)

Recommended Form of Reference to Articles

For bibliographical references to articles, the general rule is: (*a*) omit the number of the issue, and (*b*) omit the abbreviations vol. and p. or pp. whenever both occur in a reference, except in complicated references in which they are needed to clarify the meaning. (Vol. is capitalized when used with a capitalized roman numeral. In printing, Vol. is often capitalized also when used with an arabic numeral.) References should be written with "hanging indention"; punctuation should be exactly as indicated, and the spacing in accordance with the rules of typewriting. The title is placed within quotation marks, with the main words capitalized; the name of the periodical, in italics; the volume number, in roman (capital) or arabic numerals as on the periodical itself; the date is enclosed in parentheses and followed by a comma. References should be written in the following form.

```
Author's Surname, Given Names or Initials. "Title of
   Article, in quotation marks," Name of Periodical in
   italics, that is, underlined in manuscript or typescript,
   volume number in roman (capital) or arabic numerals, as
   on the periodical itself (Date of the Issue, in paren-
   theses), pagination of the article.
```

Illustrations

```
Alper, T.G. "Task-Orientation vs. Ego-Orientation in Learn-
   ing and Retention," American Journal of Psychology, LIX
   (Apr., 1946), 236-48.
Harris, Joseph, et al. "Relations of Political Scientists
   with Public Officials," American Political Science Re-
   view, XXXV (Apr., 1941), 333-43.
Holmes, Charles S. "James Thurber and the Art of Fantasy,"
   Yale Review, LV (Autumn, 1956), 17-33.
Horne, Celeste Budd. "A Geographer Looks at Russia," Cur-
   rent History, n.s., 8 (Feb., 1945), 144-9.
Kelly, Alfred H. "The Congressional Controversy over School
   Segregation, 1867-1875," American Historical Review,
   LXIV (April, 1959), 537-63.
Nosworthy, J.M. "'Macbeth' at the Globe," The Library, 5th
   Series, II (September/December, 1947), 108-18.
```

This periodical is sometimes cited with the series number before the title: 5 *The Library,* II (September/December, 1947), 108-18. The diagonal line has been used between the months because two issues were printed as one.

```
Powell, Lawrence C. "'The Western American'—An Early Cali-
   fornia Newspaper," Bibliographical Society of America,
   Papers, XXXIV (Fourth Quarter, 1940), 349-55.
Wright, David McCord. "The Future of Keynesian Econom-
   ics," American Economic Review, XXXV (June, 1945), 284-
   307.
```

If the volume number is written in arabic numerals, a colon may be used to separate the volume and page numbers. The date is sometimes placed either before or after these items, for example: (Oct. 1, 1965), 59:583–98, *or* 59:583–98 (Oct. 1, 1965). Although, in copying, this style of reference is simpler to write, the possibility of error is increased. Volume and page numbers easily become transposed because of their proximity and lack of differentiation. Moreover, in manuscript which is to be submitted to a printer the numbers should be typewritten with one space following the colon, for example: 59: 583–98. The form used in the illustrations above is to be preferred.

Title entry.—Articles for which no author is indicated are listed by title, with no change in the form of reference except the omission of the author's name. The title assumes the author's place in the arrangement.

```
"Architecture for the Ear," Time, LXIX (April 1, 1957), 76.
"The Supernatural in Scott's Poetry," Notes and Queries,
   188 (Jan. 13, 27, Feb. 24, Mar. 10, 1945), 2-8, 30-3,
   76-7, 98-101.
```

In the following reference the form is shortened because the periodical carries no volume number and the issue number is placed inconspicuously on the first page of text; the date is the only means of easily identifying the issue in which the article is published. Note the use of the abbreviation pp. in the absence of a volume number.

```
"Computers Begin to Solve the Marketing Puzzle," Business
   Week, April 17, 1965, pp. 115-38.
```

Newspaper articles.—For newspaper citations, the only items usually listed in a bibliography are the title of the newspaper and the inclusive dates of the file consulted. Titles are italicized, as are those of the periodicals. The city in which a newspaper is published should be shown, if it is not indicated in the title, for example, *Times* (London). In the text of a manuscript the title may be written more conveniently as follows: the London *Times*. As in the *Business Week* reference above, the date need not be enclosed in parentheses.

New York Times, Aug. 18-Oct. 23, 1961.
P M (New York), Mar. 1-May 16, 1944.
Times (London), Jan. 11-15, 1953.

When it is ncessary to list an important newspaper article under the name of the author or title, the reference should be written in the form of a periodical reference. Contrary to customary practice, indication of the page number is recommended. In references to the old-time four-page newspaper the pagination of an article is scarcely necessary, the date of the issue being sufficient; but when one refers to the modern thirty- to forty-page daily newspaper, often published in several sections, it is necessary to include the section and page references. It is considered desirable to give page numbers as an aid to location of the article, although it is recognized that in a newspaper like the *New York Times* the pagination may change from edition to edition. Even though the reference is not exact, an approximate reference often saves searching through an entire issue. Reference to column is desirable when the article would otherwise be less readily located on the page.

Illustrations

La Guardia, Fiorello. "Why New York Should Be the World Capital," P M (New York), Feb. 3, 1946, p. 3.
Laurence, William. "Basic Pulse Beat of Universe Seen in Particle Born Within Atom," New York Times, Feb. 2, 1947, pp. 1, 47.
"Rebuilding is Set on Maya Temples," New York Times, June 21, 1959, p. 38, col. 3.

Occasionally it is necessary to give more exact data, as in the first reference below, which cites an article published on various pages of a sixth section of a Sunday issue.

Saarinen, Aline B. "The National Gallery: Chester Dale, President and Financier-Collector..." New York Times, May 6, 1956, sec. 6, pp. 14, 30, 32-33.

Editorials should be listed by title and specially distinguished by the word "editorial" in parentheses following the title:

"The Gagged Grand Jury," (editorial), New York Times, March 5, 1962, p. 22.
"Nationalizing the Coal Industry" (editorial), Times (London), Dec. 21, 1945, p. 4.

Articles in encyclopedic works.—Articles in encyclopedias, biographical dictionaries, or similar works may be referred to in the form used for articles in periodicals, that is, the title of the article placed within quotation marks, and the title of the work written in italics. Signed articles should be listed under the author's name. Often only the author's initials are printed at the end of the article, whereas his full name will be found in a list of contributors in the first volume or at the beginning of each volume. If an article is unsigned, it should be listed by title, although some few prefer to list it under the name of the encyclopedia. The edition (if meaningful[22] and other than the first) should follow the title. If the title does not indicate that a yearbook (or supplement) of the main work is being cited, the proper designation should be made following the title (see Frankfurter and Rothwell in the illustrations below). The volume number should be given as found on the work itself (roman or arabic numerals). It is not necessary to cite the editor, publisher, or place. The date of the volume cited follows the volume number and is enclosed within parentheses; the pagination of the article is the last item. One will seldom find occasion to cite an entire encyclopedic work; the full citation is simple, however, requiring only the title, date or dates of publication, and the total number of volumes.

Illustrations

Brady, Norman C. and Thomas R. Nielsen, "Hydroponics," *Encyclopedia Americana* 14 (1964), 580-85.

"Brook Farm," *Columbia Encyclopedia,* 3d ed. (1963), p. 279.

---

[22] Many of the general encyclopedias have adopted a policy of continuous revision. Editorial staffs are constantly surveying subjects and planning revisions. Articles in any given field are revised as developments call for new treatment of the subject, even if this be annually. Numbered editions are no longer published; new printings are made each year, but only enough sets are printed to supply the year's demand. Consequently, citation of edition is often meaningless. Reference to the edition of the *Encyclopaedia Britannica* after the 14th (1929), or to the *Encyclopedia Americana* after the 2d (1918-20) should be omitted. The date is necessary, however, to identify a volume or set: *Encyclopedia Americana* (1964), XIV, 58-85. (See also the second example given above where citation of edition is necessary for identification of the *Columbia Encyclopedia*).

Forsdyke, E.J., Bernard Rackham, et al. "Pottery and Porcelain," Encyclopedia Brittanica. 18 (1945), 338-73, XLIII pls. part col.

Frankfurter, Felix. "Benjamin Nathan Cardozo," Dictionary of American Biography. Supplement Two. XXII (1959), 193-96.

Grove's Dictionary of Music and Musicians. 5th ed. (1954). 9 vols.

Nystrom, Paul H. "Retail Trade," Encyclopedia of the Social Sciences. XIII (1935), 346-55.

Rothwell, Doris P. "Living Costs and Standards," New International Yearbook. 1947, pp. 285-7.

Villard, Oswald Garrison. "Joseph Pulitzer," Dictionary of American Biography. XV (1935), 260-3.

The foregoing illustrations follow the principles outlined in this manual, and the forms given are recommended; but the form for the signed articles may be varied if an emphasis on the article itself seems important. If the form given below is adopted, it should be used throughout the bibliography; the presence of both forms would be confusing.

Saintsbury, George. The Prosody of the Nineteenth Century. In Cambridge History of English Literature, XIII (1933), 250-82.

Parts of books or sets.—When only part of a book is pertinent to the subject under discussion, this part may be cited by referring to the book in the usual manner but omitting the total paging. A citation of this kind requires a notation of the volume, section, or chapter referred to, with its inclusive paging. Finally, the title of the particular part cited is placed within quotation marks to distinguish it from the title of the book.

Venturi, Lionello. Painting and Painters. New York: Scribner, 1945. Chap. X, pp. 198-232, "Abstract and Fantastic Art."

Reference to the chapter is often given alone, without notation of the pagination of the part cited. Inclusion of the paging, however, conveys an idea of the length of the part and makes a more nearly complete reference.

One may prefer, in the reference just given, to cite first the chapter or part which is of special interest, and later to direct attention to the whole work. This may be done as follows:

Venturi, Lionello. "Abstract and Fantastic Art," in his
  Painting and Painters. New York: Scribner, 1945. Chap. X.

A book which is made up of contributions by several authors requires slightly different treatment. The author and title (in quotation marks) of the part are cited first, followed by a full reference to the book which contains the part. The volume and chapter, and the inclusive paging of the part, replace the total pagination regularly given for a book.

Mansfield, Katherine. "The Stranger," *in* W. Somerset
  Maugham, *Tellers of Tales.* New York: Doubleday, Doran,
  1939. Pp. 1178-87.

Colby, Charles C. "The Role of Shipping in the World Order," *in* Walter H.C. Laves, ed., *The Foundations of a More Stable World Order.* Harris Foundation Lectures, 1940. Chicago: University of Chicago Press, 1941. Pp. 77-103.

In the foregoing references the names of W. Somerset Maugham and Walter H. C. Laves are written in regular order. Here it is not necessary to invert the names for convenience in alphabetical arrangement.

A work in several volumes often has a collective title and additional individual titles for each volume of the set. An individual title may be indicated as part of the set in the following manner:

Gayley, Charles Mills, ed. "Dryden and His Contemporaries:
  Cowley to Farquhar." Vol. IV of *Representative English
  Comedies.* New York: Macmillan, 1936. 4 vols.

The quotation marks indicate that the volume is part of a larger work. This form is consistent with that used for parts of books; but some authors prefer to give the collective title first in the sequence, and underline both titles, as follows:

Gayley, Charles Mills, ed. *Representative English Comedies.* New York: Macmillan, 1912-36. 4 vols. Vol. IV:
  Dryden and His Contemporaries.

Needham, Joseph. *Science and Civilization in China.* Cambridge [Eng.]: University Press, 1954-65. 4 vols in 5.
  Vol. IV, pt. 2: *Mechanical Engineering,* pp. 435-546,
  "Clockwork; Six Hidden Centuries."

Bibliography 29

REFERENCES TO GOVERNMENT DOCUMENTS

Essential Points of Information

The forms of bibliographical reference used in citing government publications require an emphasis different from that of the forms already outlined. The instructions given above for books and articles, however, have certain applications here and should serve as a guide to many of the points of information required in these references. The user should note carefully the different items necessary for complete citation of (1) U. S. Congressional (legislative) publications and (2) U. S. Departmental publications.

1. GOVERNMENT AGENCY AS AUTHOR—U.S.

In a bibliography, government publications are usually listed under the name of the official agency responsible for their publication, that is, the "corporate author" (cf. pp. 17–18). In this form of reference, one should indicate first the name of the country, state, municipality, etc. (e.g., United States, usually abbreviated U. S., Illinois, Boston); second, the major division of the government (e.g., Congress, Legislature, Dept. of Commerce, Bureau of Fisheries, City Council); and then further subdivisions, if necessary (e.g., Senate. Committee on Manufactures; Legislature. Assembly).

In a bibliography, a long dash, made in typing by striking four hyphens in succession (— — — —), in printing by the use of a 3-em dash (———), may be used to indicate an exact repetition of a corporate author's name, if the repetition occurs on the same page or on any two pages which would lie open to the reader at the same time. The first long dash, when used for repetition of a reference to any U. S. document, always includes U. S., which is considered to be part of the corporate entry, e.g., U. S. Congress, U. S. Bureau of American Ethnology, U. S. Public Health Service. If a second subdivision is repeated, it must be represented by a second long dash (——— ———), e.g., U. S. Congress. Senate, *or*, U. S. Dept. of State. Division of Cultural Cooperation. Occasionally, three long dashes (——— ——— ———) will be needed, when a third subdivision of the corporate author is repeated, e.g., U. S.

Congress. Senate. Committee on Interstate Commerce. See also the repetitions in the illustrations under League of Nations and United Nations documents, and compare personal author repetition, page 16). The following examples show various types of government agencies as corporate author:

```
U.S. Congress. House. Committee on Ways and Means.
---- Senate. Committee on Foreign Relations.
U.S. Office of Education.
---- Statistical Division
U.S. Bureau of Foreign and Domestic Commerce.
Massachusetts. Dept. of Corporations and Taxation.
---- ---- Income Tax Division.
California. Legislature. Senate.
---- State Planning Board.
```

Examples of the correct form of the name of a federal government agency as corporate author may be found in a library card catalog, the *Monthly Catalog of United States Government Publications*,[23] *United States Government Organization Manual*,[24] or bibliographies of government publications. The *Monthly Checklist of State Publications*[25] lists the names of state agencies.

2. PERSONAL AUTHOR

A government publication showing a personal author on the title page should be listed in a bibliography under the name of the official issuing agency—the corporate author. The personal author is shown by the addition of his name after the title:

```
U.S. Congress. Senate. The Dumbarton Oaks Proposals and the
    League of Nations Covenant, by Herbert Wright. 79th
    Cong., 1st sess., S. Doc. 33. Wash.: G.P.O., 1945. 38 pp.
U.S. Department of State. Memorandum on the Postwar Inter-
    national Information Program of the United States, by
    Arthur W. Macmahon. Dept. of State Pub. no. 2438. Wash-
    ington: G.P.O., 1945. 135 pp.
```

The form twice illustrated above is preferred, but the publications

---

[23] (Washington: 1895–   ). An index to government publications which has been expanded to include features of the discontinued *U.S. Documents Catalog* (Washington: 1893–1940). See Laurence F. Schmeckebier and Roy B. Eastin, *Government Publications and Their Use* (rev. ed.; Washington: Brookings Institution, [1961]), pp. 38–40.

[24] U.S. General Services Administration (Washington: G.P.O., 1935–   ).

[25] U.S. Library of Congress (Washington: G.P.O., 1910–   ).

may be listed under the name of the personal author; the official aspect of the publication is then shown in the notation of the series and number, as follows:

Wright, Herbert. <u>The Dumbarton Oaks Proposals and the League of Nations Covenant.</u> U.S. 79th Cong., 1st sess., S. Doc. 33. Washington: Government Printing Office, 1945. 38 pp.

Macmahon, Arthur W. <u>Memorandum on the Postwar International Information Program of the United States.</u> U.S. Dept. of State Pub., no. 2438. Washington: Government Printing Office, 1945. 135 pp.

The form of author entry adopted should remain consistent throughout the bibliography, whether it is personal or corporate.

3. TITLE

The titles of government documents are often long and involved, especially the United States Congressional reports and documents, which are cited frequently. Many authors abbreviate the titles and indicate the omissions by some form of ellipsis (e.g., ...); others use a shortened title without indication of the omissions. The safest practice is to note the title in full. If it is cumbersome or redundant, one can reduce it intelligently and indicate omissions by the use of three dots. It is recommended that the titles of documents which form complete works in themselves be italicized (underlined in manuscript or typescript), and that those titles which are no more than parts of individual publications be placed within quotation marks (cf. "Alternate Form," p. 37).

4. IMPORTANCE OF SERIES AND NUMBER

Many government publications can only be identified and located in libraries by reference to the series and number. Consequently, the series notation often becomes the most important item in the reference.

5. EDITION

If the edition of a government document is not the first, it should be noted in the form recommended for books, i.e., the edition should appear as the item which precedes the imprint.

6. IMPRINT

In referring to United States departmental publications, it is increasingly important to give the full imprint (place: publisher, date). The Government Printing Office issues annually more than a billion copies of government publications which will constitute the bulk of materials ordinarily cited. They will present no problem. Prior to World War II, Washington, and the Government Printing Office appeared to be such obvious imprint items that they were often omitted. However, owing to the postwar expansion of government services, and of research and development, a rapidly accelerating number of significant publications were issued by individual government agencies themselves (*not* by the G.P.O.), and also through commercial printing agreements. Nearly all administrative agencies now follow this practice to some extent. Consequently, this category of government publication, especially in the fields of science and technology, presents a variety of imprints which require full identification. It must not be taken for granted that departmental publications are issued by the Government Printing Office.

Congressional (legislative) publications are issued by the Government Printing Office, which should be noted in the imprint. References to the Congressional series often omit the date, since the number of the Congress and of the session are thought to be sufficient identification. This confusing practice should be avoided because the date and the number of a Congress are not readily associated in the minds of many people.

7. SERIAL NUMBER

United States government publications are issued separately, but are finally bound as a set the volumes of which are numbered consecutively. This set is known as the "serial set." The Government Printing Office assigns to each volume its individual number, known as the serial number. The complete set contains both Congressional publications and Departmental publications. Inclusion of the serial number in a reference is optional, but its presence often saves searching through various indexes if the serial number is necessary for the location of a volume in a library. When citing current or recent documents, however, it is not always possible to include the serial number, since there is a lapse of time (often five years) between the first publication of a document and its incorporation in the serial set. The serial number, if cited, should be the last item in the reference.

Recommended Form of Reference to Documents

U.S. Congress. Senate. **Reference Manual of Government Corporations . . . as of June 30, 1945.** 79th Cong., 1st sess., S. Doc. 86. Washington: G.P.O., 1945. 526 pp. Serial no. 10947.

Bibliographical references to U. S. government documents should contain the essential items of information discussed above (under centered headings 1 to 7). Variations in the form and sequence of these items in different types of government publications are discussed and illustrated individually in subsections 1 to 6 below. Publications of foreign governments and international bodies are treated in subsections 7 to 9.

1. U. S. CONGRESSIONAL PUBLICATIONS

Publications issued by either of the two Houses of the United States Congress (House of Representatives or Senate) include Bills, Resolutions, Reports (of committees), Documents, and other series, each numbered consecutively throughout a Congress (two years). In referring to one of these publications, the series should be noted by an indication of (a) the number of the Congress, (b) the session, (c) the House (i.e., House of Representatives or Senate), (d) the type of publication (i.e., Bill, Resolution, Report, or Document), and (e) its number. Notation of the series and number of Congressional publications should be shortened by using abbreviations:

79th Cong., 1st sess., H.R. 241.
    For House Bill 241. Do not confuse with H. Res.

58th Cong., 2d sess., S. 162.
    For Senate Bill 162.

71st Cong., 3d sess., H. Res. 1.
    For House Resolution 1.

62d Cong., 1st sess., S. Res. 19.
   For Senate Resolution 19.
79th Cong., 2d sess., H. Rept. 228.
   For House Report 228.
80th Cong., 1st sess., S. Rept. 123.
   For Senate Report 123.
72d Cong., 2d sess., H. Doc. 438, pt. 2.
   For House Document 438, part 2.
56th Cong., 2d sess., S. Doc. 289.
   For Senate Document 289.

Reference may be made in the same manner to House or Senate Concurrent Resolutions (Con. Res.), Joint Resolutions (J. Res.), Executive Documents (Ex. Doc.), and Miscellaneous Documents (Misc. Doc.).

When considering proposed legislation (bills), Congressional committees often find it necessary to summon witnesses for expert testimony, or to make sure that differences of opinion are taken into account in the course of deliberations. In references, such committee hearings are indicated by the word "Hearings," which follows the title and is separated from it by a period. Hearings are unnumbered, and in order to avoid confusion in their identification it is necessary to include the number of the Senate or House bill, resolution, etc., with which they are concerned. The series note is the same as for the Congressional publications above.

Illustrations

Bills and Resolutions

U.S. Congress. Senate. An Act for the Development and Control of Atomic Energy. 79th Cong., 1st sess., S. 1717. Washington: G.P.O., 1945. 26 pp.

U.S. Congress. House. A Resolution for the Consideration of S. 1717 for the Development and Control of Atomic Energy. 79th Cong., 2d sess., H. Res. 708. Washington: G.P.O., 1946. 2 pp.

Hearings

U.S. Congress. Senate. Special Committee on Atomic Energy. Atomic Energy Act of 1946. Hearings, 79th Cong., 2d sess., on S. 1717. Washington: G.P.O., 1946. 573 pp.

---- House. Military Affairs Committee. **Atomic Energy.**
Hearings, 79th Cong., 2d sess., on S. 1717. Washington:
G.P.O., 1946. 68 pp.

Reports

U.S. Congress. Senate. Special Committee on Atomic Energy.
Atomic Energy Act of 1946. 79th Cong., 2d sess., S. Rept.
1211 to accompany S. 1717. Washington: G.P.O., 1946.
125 pp.

---- House. Military Affairs Committee. **Atomic Energy Act
of 1946.** 79th Cong., 2d sess., H. Rept. 2478 to accompany
S. 1717. Washington: G.P.O., 1946. 21 pp.

Documents

U.S. Congress. Senate. **Proposals for Improving the Patent
System.** Subcommittee . . . Committee on the Judiciary.
Study no. 1, by Vannevar Bush. 84th Cong., 2d sess., S.
Doc. 3724. Washington: G.P.O., 1957. 30 pp.

---- House. **Annual Report of the National Science Foundation for the Fiscal Year, 1962.** 88th Cong., 1st sess., H.
Doc. 4714. Washington: G.P.O., 1963. 368 pp.

2. U. S. DEPARTMENTAL PUBLICATIONS

Departmental publications include any publication issued by a department, bureau, office, authority, or other agency of the administrative branch of the government. Most departments issue bulletins and other series, in addition to annual reports and miscellaneous works. In bibliographical citations, a bureau is treated as a major subdivision of the United States government; it should not be listed as a subdivision of a department (e.g., U. S. Internal Revenue Service, *not* U. S. Treasury Dept. Internal Revenue Service). A division, however, is considered a subdivision of the larger agency, i.e., department, bureau, office, etc., and should be listed subordinately (see the fifth illustration below). The references should be written in the following form:

The Issuing Agency as Author. **Title of the Document in Italics.** Personal Author, if any. Series and number. Edition, if necessary. Place: Publisher, Date. Main pagination or total volumes. Illustrations, etc., if pertinent. Serial number, optional.

Illustrations

U.S. Atomic Energy Commission. **Nuclear Reactors Built, Building, or Planned as of Dec. 31, 1960.** TID 8200 3d rev. Oak Ridge, Tenn.: A.E.C., 1961. 26 pp. Processed.

U.S. Bureau of Foreign and Domestic Commerce. Small Retail Store Mortality, by William T. Hicks and Walter F. Crowder. Economic Series, no. 22. Washington: G.P.O., 1943. 43 pp.

U.S. Business and Defense Services Administration. Soviet Research on Corrosion of Special Alloys. Washington: Office of Technical Services, 1961. 143 pp. Processed.

U.S. Civil Aeronautics Administration. Pilot's Radio Handbook, by Tom Dodson. CAA Technical Manual, no. 102. Washington: G.P.O., 1954. 122 pp.

U.S. Department of State. Division of Cultural Cooperation. American Republics Branch. Exchange of Specialists and Distinguished Leaders in the Western Hemisphere, by Francis J. Colligan. Inter-American Series, no. 27. Washington: G.P.O., 1946. 14 pp.

U.S. Health, Education, and Welfare Department. Health Insurance for Aged Persons: Report Submitted to House of Representatives Committee on Ways and Means by Secretary . . . July 24, 1961. Washington: G.P.O., 1961. 119 pp. Processed.

U.S. Inland Waterways Corporation. Annual Report . . . to the Secretary of Commerce, Calendar Year 1943. Washington: G.P.O., 1944. 31 pp.

U.S. National Aeronautics and Space Administration. Study in Cometary Astrophysics, by Vincent J. De Carlo. NASA Contractor Report Series 14772. Falls Church, Va.: Melpar, Inc., 1964. 22 pp.

---- Tides in Atmosphere of Earth and Mars, by Richard A. Craig. NASA Contractor Report Series no. 18722. Bedford, Mass.: Geophysics Corp. of America, 1964. 42 pp.

U.S. President. Economic Report of the President Transmitted to Congress January 1966 Together with the Annual Report of the Council of Economic Advisers. 89th Cong., 2d sess., H. Doc. 348. Washington: G.P.O., 1966. 306 pp.

May be cited also as a Congressional publication.

U.S. Securities and Exchange Commission. Directory of Companies filing Annual Report . . . Under Securities and Exchange Act of 1934. Washington: G.P.O., 1964. 203 pp.

U.S. Technical Services Office. Government-owned Inventions Available for License, Accumulated Jan-Dec. 1963. Patent Abstract Series, no. 15135. Washington: T.S.O., 1964. 108 pp.

If any of the illustrations above are cited by the personal author, the agency responsible for the publication (corporate author) may then be indicated preceding the series note. (For examples of documents cited in both forms, see pp. 30–31.) Note that processed documents (reproduced by duplicating processes other than ordinary printing) are so identified at the end of the reference.

In the periodic reorganization of the administrative branch of the government, agencies are combined, transferred from one department to another, and sometimes eliminated and the functions transferred elsewhere. In references to government documents, one should always use the name of the agency current at the time of publication of the reference being cited.

Alternate form of document reference.—An alternate form of reference places the title of the document within quotation marks and uses italics for the series notation, as, for example:

U.S. Congress. House. Military Affairs Committee. "Atomic Energy Act of 1946." *79th Cong., 2d sess., H. Rept. 2478 to accompany S. 1717.* Washington: G.P.O., 1946. 21 pp.

U.S. Department of State. Division of Cultural Cooperation. American Republics Branch. "Exchange of Specialists and Distinguished Leaders in the Western Hemisphere," by Francis J. Colligan. *Inter-American Series,* no. 27. Washington: G.P.O., 1946. 14 pp.

This form is not recommended, because it does not differentiate between document references proper and the government periodical references described below. If the alternate form is adopted, it should be used exclusively throughout the manuscript.

Periodicals published by the government.—Articles in periodical publications (e.g., *Monthly Labor Review, School Life, The Department of State Bulletin,* etc.) which appear at fixed intervals and contain articles that can be identified by volume, date, and page are cited in the same manner as other periodical articles (see pp. 20 ff.).

Wilcox, Francis O. "The United States and the Challenge of the Underdeveloped Areas of the World," *The Department of State Bulletin,* XL (May 25, 1959), 750-58.

3. CONGRESSIONAL RECORD

While Congress is in session, the *Congressional Record* (verbatim record of debates and remarks) is published in daily and semimonthly issues. Later, when the permanent bound edition is published, it is repaged. Consequently, a confusion in the page references often results. References should be made to the bound edition. If this is impossible, the reference should indicate that the daily or semimonthly unbound edition is being cited.

The *Congressional Record* should be listed in the bibliography under title, with notation of the Congress, session, volume, and inclusive dates. Specific page references are found in the footnotes (see pp. 99–100).

Congressional Record. 80th Cong., 2d sess., vol. 94, pt. 3 (Mar. 22-Apr. 7, 1948).

A fuller form of reference gives the inclusive pages, and may be written:

Congressional Record. 80th Cong., 2d sess., 94: 3 (Mar. 22-Apr. 7, 1948), 6085-7038.

Some authors list only the title in the bibliography, but that kind of general reference is seldom justified. If the references become numerous and cumbersome, they may be simplified, but the inclusive dates or period covered should always be included, as, for example:

Congressional Record. 80th Cong., 2d sess., Mar. 22-Apr. 7, 1948.

4. U. S. LAWS, STATUTES, ETC.

Ordinarily, individual laws (enactments of Congress, approved by the President) are not listed in the bibliography; the specific citation is made in the footnotes (see pp. 100–101). To indicate that laws have been used in the preparation of a paper, one should list by title the official set of laws consulted, and, where necessary, indicate the date and edition, particularly if an annotated or other special edition was used.

Code of Laws of the United States (1940 ed.).
 Or:
United States Code (1940 ed.).
 Or:
United States Code, Supp. IV (1945).
Internal Revenue Code, (1954).
U.S. Code Annotated (Suppl. 1938).
U.S. Revised Statutes.
U.S. Statutes at Large.

 For this last, the volume numbers may be given.

If it is necessary to cite a Public Act in the bibliography, this should be done more specifically. When a bill has been passed by both House and Senate, and has been signed by the President, it becomes a law, i.e., a Public Act. Public Acts (called, in their first printed form, "slip laws") are issued currently while Congress is in session. They later appear in the bound volumes of the *U.S. Statutes at Large*. A reference to a Public Act should give the number of the act, the number and session of the Congress enacting it, and, in parentheses, the date. Often, the name of the act accompanies the citation; this is not necessary, but if it is given, it should be enclosed within quotation marks.

<u>Public</u> <u>Law</u> no. 585, 79th Cong., 2d sess. (Aug. 1, 1946).
   "Atomic Energy Act of 1946" may be added after the date.

This citation should be made to the *Statutes at Large* as soon as the Public Act has been incorporated in that compilation, and would probably be placed in a footnote (see p. 101). The same instructions apply to the citation of Private Acts.

The Constitution is listed in the bibliography simply as:

<u>Constitution</u> <u>of</u> <u>the</u> <u>United</u> <u>States</u> <u>of</u> <u>America.</u>

If desirable, it may be grouped with other entries under United States as follows:

<u>U.S.</u> <u>Constitution.</u>

5. U. S. SUPREME COURT, ETC.

The first ninety-one volumes of the reports of the United States Supreme Court decisions were published under the names of the official reporters (e.g., Cranch, Wheaton, etc.), but since 1875 the reporters' names have been disregarded and the reports have been published under the title *United States Supreme Court Reports*. The early volumes have been renumbered so that they now run consecutively from volume 1 (1789) to date. If one wishes to list Supreme Court reports in a bibliography, it is sufficient to give the following reference:

<u>U.S.</u> <u>Supreme</u> <u>Court</u> <u>Reports.</u>

If the *Supreme Court Reporter, Lawyers' Edition,* or the advance sheets were used, this fact should be indicated. The citation of individual decisions is made specifically in the footnotes (see p. 103).

A specific citation of reports of the lower federal courts, i.e., United States Circuit Courts, District Courts, etc., will seldom be made except in footnotes. The reference in the bibliography may be made by giving the title of the compilation containing the report; for example: *Federal Reporter, Second Series.* Reports of administrative bodies and special courts may be referred to in the same manner: *U. S. Interstate Commerce Commission Valuation Reports; U. S. Court of Customs and Patent Appeals Reports* (pt. 1, Customs); *Federal Register.*

6. STATE AND MUNICIPAL DOCUMENTS

References to state and municipal legislative and administrative documents are based upon the same principles as the references for United States documents. The citation should include the name of the state or municipality; the major subdivision of the government (Legislature. Assembly; Department of Agriculture; City Council; Board of Education; etc.); the title of the document (in italics); series (if any); imprint; main pagination.

Illustrations

Baltimore. Commission on Health. Guarding the Health of Baltimore: A Summary of the One Hundred and Forty-second Annual Report . . . 1956. Baltimore: 1957. 65 pp.

California. Legislature. Joint Committee on Water Problems. Preliminary Report to the Legislature, 1948 Regular Session, on Water Problems of the State of California. Sacramento: 1948. 77 pp.

Detroit. Capital Improvement Program Committee. A Six-Year Reserve Capital Improvement Program for the City of Detroit, 1943-49. Detroit: 1943. 83 pp.

Illinois. Governor. Message of Governor William G. Stratton to the 71st Assembly, Wednesday, January 7, 1959. [Springfield, 1959]. 23 pp.

Indiana. Board of Tax Commissioners. Manual of Real Estate Assessment for the State of Indiana. Indianapolis: 1941. 48 pp.

New York (City). Mayor's Committee for World Fashion Center. The World Fashion Center, New York City's Post War Business Project, No. 1: A Report . . . New York: 1944. 221 pp. Illus., plans.

State and municipal documents are seldom issued by official printing offices, and the various printers' names are not particularly significant bibliographically. Indication of the place and date of publication may suffice. If the form of entry (corporate author) presents difficulties, one may consult the official manual or bluebook of the state concerned, or the *Monthly Checklist of State Publications* which has been issued by the Library of Congress since 1910.*

State laws.—The statutes of the individual states (enactments of the legislatures) are referred to variously as laws, acts, statutes, or session laws, and should be listed in the bibliography under the title used on the publication itself. The reference should include also the year or session of the legislature. Ordinarily, each act (law) is numbered as a chapter, and the divisions of the act are numbered as sections, but specific citation to these subdivisions is given in the footnotes (see p. 101).

Compilations of statutes are called "codes." Codes may be a restatement of the general or permanent laws of the state, or may be a collection pertaining only to a branch of the law, such as civil, penal, political, criminal, or education codes. The reference in the bibliography is usually a general one, listing only the statutes or codes used, with the dates in parentheses, and, when advisable, the name of the editor. If an annotated or revised edition of statutes or codes is used, this should be indicated as shown below, but notation of imprint, pagination, or volumes is unnecessary. The specific citation is made in the footnotes (see p. 102).

<u>California</u> <u>Penal</u> <u>Code.</u>
    California codes are not cited by editor or date.
<u>Indiana</u> <u>Statutes</u> <u>Annotated</u> (Burns, 1933).
<u>Iowa</u> <u>Code</u> (1946).
<u>Kansas</u> <u>Laws</u> (1945).
<u>Nevada</u> <u>Statutes</u> (1943).
<u>South</u> <u>Carolina</u> <u>Acts</u> (1946).

A state constitution is listed in the bibliography in the following simple style:

---

* The data given in the text suffice to identify this publication; consequently, a footnote reference is unnecessary.

Arizona Constitution
If the original Constitution has been superseded, the date of the Constitution used should be given:
New York Constitution (1939).

State court reports.—Decisions of state courts are reported officially in individual "State Reports" in much the same manner as are those of the United States Supreme Court. These decisions are reported continuously in the National Reporter System, which groups the states more or less geographically and publishes one "Reporter" for each section. Citation of individual cases is made in the footnotes (see p. 104); only the title of the compilation need be given in the bibliography, and this is sometimes omitted.
California Reports.
Northeastern Reporter.

City charters and ordinances.—The laws defining the government of a municipality may be found scattered through the statutes enacted by the state legislature, or they may be in the form of a special instrument. If they are scattered through the statutes, bibliographical reference to the statutes is made in the manner described on page 41 above. Specific footnote citation to the individual statutes is described on page 102 below. If the scattered statutes have been collected in a single publication, bibliographical reference to the collection is made in the form illustrated just below. Likewise, the special instrument (and its amendments) may be cited either as it first appeared in the statutes or as a separate publication. "Home rule" charters are cited in the manner illustrated. The form of entry which lists the name of the city first in the bibliography allows an alphabetical arrangement by city.

New York (City). Charter. The New York City Charter Adopted by Referendum November 3, 1936 . . . as Amended to November 1, 1942. Annotated. Albany: 1943. 2556, [88] pp.
―――― ―――― Proposed Charter for the City of New York . . . . New York: Charter Revision Commission, [1961]. 155 pp.

Neither the publication nor the citation of city ordinances (laws) has been standardized. Ordinances are printed in newspapers. They may be printed separately, but usually they are available only in the codes which are made up of laws concerning one subject or related subjects. The name of the city should be brought out clearly, and the title of the ordinance should be given, together with enough other data to identify and locate the law. Many authors prefer to list ordinances by title, but the form below is recommended.

Chicago. Ordinances. Chicago Zoning Ordinance . . . Passed by the City Council on May 29, 1957, as Amended Nov. 25, 1958. Chicago: Index Pub. Corp., 1959. 56A, 248B, 12C pp.

Portland, Ore. Ordinances. License and Business Code of the City of Portland, Oregon . . . Edited and compiled by the Bureau of Municipal Research and Service, and the University of Oregon cooperating with the League of Oregon Cities . . . [Portland]: 1941. 1 vol. loose-leaf.

7. BRITISH GOVERNMENT DOCUMENTS, ETC.

British Government documents are roughly classified as Parliamentary publications and non-Parliamentary publications. They are printed and distributed officially by Her Majesty's Stationery Office, which corresponds to the United States Government Printing Office.

Parliamentary publications.—Parliamentary publications handled by H.M. Stationery Office are collected and published annually in a complicated series of volumes referred to as "Parliamentary Papers," "Blue Books," or "Sessional Papers." The Parliamentary publications contain House of Commons Papers (abbreviated H.C.) and House of Lords Papers (H.L.), which comprise documents presented to the respective Houses. They include bills and amendments, reports of joint, select, and standing committees, reports of commissioners, records of financial negotiations, in fact any business actually before Parliament or matters about which members require information. Documents of this class are distributed free to members of Parliament. In order to reduce the rapidly increasing quantities of publications in this class, and to save expense, series are transferred at intervals to the non-Parliamentary category; consequently, there is no longer a clear-cut division between these two classes of documents.

The annual series of House of Commons Papers is published in a series of many volumes, arranged in the following groups: (1) Public Bills, (2) Reports of Committees, (3) Reports of Commissions, etc., and (4) Accounts and Papers. Each of these groups may contain five or more volumes, and each group carries its own sequence of arabic numerals. In addition, the annual series carries its own sequence of volume numbers in consecutive roman numerals. These two sets of numbers differentiate the group volume numbers from the series volume numbers (e.g., Parliamentary Publications, 1943–44, Vol. XXIII (Public Bills, vol. 4). For complete identification of references, group and series numbers should be included.

Command Papers are, theoretically, presented to Parliament by command of the Sovereign. Since 1900 they are found only in the House of Commons Papers. The more important reports and publications are issued in this series, which constitutes the largest single category in the Commons Papers each year. Three sets of symbols, in the lower left-hand corner of the document, roughly classify this series by date: C. 1 to C. 9550, from 1870 to 1899; Cd. 1 to Cd. 9239, from 1900 to 1918; Cmd. 1 to Cmd. 9889, from 1919 to 1956; and Cmnd. 1–, from 1957 to the present.[26] The sequence of numbers and the initials are changed just before the number reaches five digits. Since these are bound with the House of Commons Papers, for complete identification it is necessary to give the Command Paper number, e.g., Cmd. 2014, in addition to the date and volume number of the Parliamentary publications, e.g., Parliamentary Publications, 1937–38, Vol. XXXIV (Committee Reports, vol. 7), Cmd. 5885.[27] References are often given with the Command Paper number as the only means of identification, but references in that form only make difficulties when one is attempting to locate documents in a library.

Most of the Parliamentary publications which the student will find occasion to cite are contained in the House of Commons Papers, which constitute the more important and inclusive series.

---

[26] Percy Ford and Grace Ford, *A Guide to Parliamentary Papers: What They Are; How to Find Them; How to Use Them* (Oxford: Blackwell, 1955), pp. 24–26.

[27] *Ibid.*, pp. 45–48.

Bibliography

The House of Lords Papers are bound annually on the same general plan, but they amount to comparatively few volumes, containing bills, *Minutes of Proceedings,* votes, etc.

The items to be included in bibliographical references correspond to those outlined under United States documents (pp. 29 ff.), i.e., corporate author and necessary identification of governmental subdivision, title, edition, series, and imprint.

Illustrations

Great Britain. Forestry Commission. Post-war Policy Private Woodlands. Supplementary Report. Parliamentary Publications, 1943-44, Vol. III (Commissioners' Reports, vol. 1). Cmd. 6500 London: H.M.S.O., 1945. 11 pp.

---- Ministry of Housing and Local Government. Housing Return for England and Wales, 30th June 1960. Parliamentary Publications, 1959-60, Vol. XXVII (Accounts and Papers, vol. 6), Cmnd. 1119. London: H.M.S.O., [1960]. 33 pp. incl. statis. tables.

---- Ministry of Labour and National Service. Report on Post-war Organisation of Private Domestic Economy. Parliamentary Publications, 1944-45, Vol. V (Commissioners' Reports, vol. 2). Cmd. 6650. London: H.M.S.O., 1945. 26 pp.

---- Parliament. Joint Select Committee of House of Lords and House of Commons Appointed to Enquire into Accommodations in the Palace of Westminster, Report . . . . Parliamentary Publications, 1943-44, Vol. II (Committee Reports, vol. 1), H.L. 50; H.C. 116. London: H.M.S.O., 1944. 60 pp.

---- ---- House of Commons. Agreements . . . Regarding Financial Assistance to Czecho-Slovakia. . . Jan. 27, 1939. Parliamentary Publications, 1938-39, Vol. XXVII (Accounts and Papers, vol. 12), Cmd. 5933. London: H.M.S.O., 1940. 15 pp.

May be cited also as *Treaty Series,* no. 9 (1939).

---- ---- ---- A Bill to Prohibit the Hunting with Hounds of Deer; to Provide for the Control of Deer by Approved Methods; and for Purposes Connected Therewith. Parliamentary Publications, 1959-60, Vol. III (Public Bills, vol.3), H.C.B. 125. London: H.M.S.O., 1960. 2pp.

---- ---- ---- Finance Accounts of the United Kingdom for the Fiscal Year, 1961-62. Parliamentary Publications, Vol. XXV (Accounts and Papers, vol. 1), H.C. 217. London: H.M.S.O., 1962. 71 pp.

---- ---- House of Lords. **A Bill** [**as amended by joint select committee**] **Intituled an Act to Consolidate with Amendments Certain Enactments Relating to Local Government in London.** Parliamentary Publications, 1938-39, House of Lords Sessional Papers, Vol. III, H.L. 103. London: H.M.S.O., 1939. 170 pp.

---- Privy Council. Committee for Medical Research. **Report of the Medical Research Council for the Year, 1938-39.** Parliamentary Publications, 1939-40, Vol. IV (Commissioners' Reports, vol. 1) Cmd. 6163. London: H.M.S.O., 1940. 171, xv pp.

Non-Parliamentary publications.—The non-Parliamentary publications consist of documents of general public interest which are published by authority of the departments, and not by command. These publications include *Statutory Rules and Orders* which govern the activities of the various departments; *British and Foreign State Papers; Herstlett's Treaties;* publications of state archives under the direction of the Historical Manuscripts Commission; Foreign Office *Peace Handbooks; Board of Trade Journal;* and publications in science, technology, public health, fine arts, etc. They are comparable to U. S. Departmental publications, and in bibliographical references are arranged under the department or other branch of government responsible for their contents. The proper form of entry may be determined by consulting the library card catalog or the *Government Publications Catalogue* published by H.M. Stationery Office since 1922, and the supplementary *Monthly Lists.* The Historical Manuscripts Commission's *Guide to the Reports and Collections of Manuscripts . . . ,* Parts 1 (1914) and 2 (1935), includes lists of "Reports Referred to" which give brief titles greatly simplifying references to volumes of the Commission's early publications. Until 1899 each *Report* was followed by an elaborate set of appendixes. An analytic chart presenting a "list of Reports, Appendixes, Etc.," will be found in *The Sources . . . of English History,* by Charles Gross.[28] The items in a bibliographical reference to an non-Parliamentary publication are the same as those described under U. S. Departmental publications, pages 35 ff. above. In references to British documents it is advisable to indicate the publisher.

---

[28] (2d ed.; London: Longmans, Green, 1915), App. B, pp. 692–7.

Illustrations

Great Britain. Board of Trade. Journal of the Commissioners for Trade and Plantations Preserved in the Public Record Office. London: H.M.S.O., 1920-38. 14 vols.
---- Colonial Office. Report by His Majesty's Government . . . to the Council of the League of Nations on the Administration of Palestine and Trans-Jordan, 1938. Colonial, no. 166. London: H.M.S.O., 1939. 422 pp. Maps.
---- Foreign Office. Historical Section. International Congresses, by Ernest Satow. Peace Handbooks, Vol. XXIII, no. 151. London: H.M.S.O., 1920. 168 pp.
---- Historical Manuscripts Commission. Calendar of the Stuart Papers . . . Preserved at Windsor Castle. London: H.M.S.O., 1902-23. 7 vols.
---- Ministry of Transport. Traffic in Towns; A Study of the Long Term Problems of Traffic in Urban Areas . . . London: H.M.S.O., 1963. 223 pp. maps.
Victoria and Albert Museum. English Silversmiths' Work. Civil and Domestic: An Introduction, by C. Oman. London: H.M.S.O., 1966. 20 pp. 193 pls.

The examples by Satow and Oman may be entered under personal author; see pages 30–31 and cf. page 91.

Parliamentary Debates.—The *Parliamentary Debates* are published under the control of the House of Commons, and are printed by H.M. Stationery Office. Originally, the reporting and printing of debates was a private venture of the Hansard family, and the name Hansard has been retained on the title page (in parentheses) and is often used conventionally in references. These are cited variously: to 1803 as *Hansard, Parliamentary History;* from 1803 to 1908 as *Parliamentary Debates,* with or without the name of Hansard; and beginning with the fifth series in 1909, citations are made individually to *House of Commons Debates* and *House of Lords Debates.* Specific references are made in the footnotes (see p. 100). Often only the title is listed in the bibliography. It is recommended that the bibliographical reference include an indication of the series used, the specific volumes consulted, and the dates covered.

Hansard Parliamentary History, Vol. XVI (1765-71).
Parliamentary Debates. 4th series, vols. 58-9 (1898-99).
House of Commons Debates. 5th series, vol. 419 (1945-46).
House of Lords Debates. 5th series, vol. 138 (1945-46).

English statutes.—In addition to the ordinary necessities for statute citations, as we are familiar with them, there are further occasions when English statute citation is required.[29] Since the powers of the government are not formally defined in a written constitution, it is often necessary to cite the fundamental laws enacted by Parliament which formulate and regulate the rules of common law, machinery of government, etc. "Slip laws" are printed before the bound volumes of statutes are compiled, but these "slip laws" are generally unavailable in American libraries. In a bibliography it will seldom be necessary to list more than the title of the compilation containing the law; specific citations are made in the footnotes (see p. 102).

Chitty's Statutes of Practical Utility.
Great Britain Statutes at Large.
Law Report Statutes.
Statutes of the Realm.

English court reports.—The earliest systematic reports of English cases are recorded in the *Year Books* (1292–1534), so named because the cases were compiled by regnal year (i.e., year of reign). Later court reporting was a private venture, and the quality and degree of completeness of the reports depended upon the talents and efforts of individual reporters. These reports are cited by the name of the reporter and pertain usually to a special branch of the law or to a single court, e.g., *Brown's Chancery Cases, Moore's King's Bench, etc.* The majority of cases of this period have been reprinted in the collection *English Reports—Full Reprints* (also called *English Reprints* and *English Reports Reprints*), which are often cited in preference to the originals. Centralized reporting of English court decisions was initiated by the government with the official publication of *Law Reports* in 1865. This compilation is divided into several series, each of which gives the decisions of one of the various courts, e.g., Common Pleas, Exchequer, House of Lords (including Privy Council), Chancery Courts, etc. These *Reports* are not a complete collection of all decisions—the selec-

---

[29] Percy H. Winfield, *Chief Sources of English Legal History* (Cambridge: Harvard University Press, 1925), 374 pp. This is a full treatment of both the early statutes and court reports.

tions reflect the editor's choice—but the compilation is more than adequate for most purposes. Paralleling and often supplementing the official reports are other editions, such as the *Times Law Reports, Law Reports,* or *Law Journal Reports.* Specific citation of cases is given in the footnotes (see pp. 104–105). The compilations are listed in the bibliography by title.

8. OTHER FOREIGN GOVERNMENT DOCUMENTS

References to other foreign government publications may be adapted to the forms outlined for United States and British documents. Legislative publications and debates, administrative publications, statutes, etc., follow similar patterns, and the same items within the reference will need consideration, e.g., the corporate author with proper identification of government subdivisions, personal author, title, series, imprint, and pagination, which are discussed on pages 29 ff. The *List of Serial Publications of Foreign Governments, 1814–1931,* edited by Winifred Gregory,[30] is an invaluable aid in determining the entries. The following examples are given for the purpose of illustrating resemblance to references of other government documents which are discussed above in greater detail.

Illustrations

Argentine Republic. Consejo Federal de Inversiones. Zona Latinoamerica de Libre Commercio: Reseña Juridíca. Colección Divulgación, no. 1. Buenos Aires: 1963. 110 pp.
---- Junta Reguladora de Vinos. . . . Censo de Viñedos, Año 1936. Actualizado de Acuerdo a las Extirpaciones Realizades en los Años 1936 a 1938 en Cumplimiento de las Leyes 12.137 y 12.355. Buenos Aires: Republica Argentina, 1938. 229 pp. Maps.
France. Assemblée Nationale. Chambre des Députés. Commission d'Enquête Chargée de Rechercher touts es Responsabilités Politiques et Administratives Encourues depuis l'Origine des Affaires Stavisky . . . Rapport Générale Fait au Nom de la Commission . . . Chambre de Deputes, 15 Legis. sess., 1935, no. 4886, Annexe au Procès-verbal, 7 mars 1935. Paris: Imprimerie de la Chambre de Députés, 1935. 827 pp.

---

[30] (New York: Wilson, 1932).

France. Commission pour la Publication des Documents Relatifs aux Origines de la Guerre . . . Documents Diplomatiques Français, 1871-1914. 3e sér., Vol. XI, 24 juillet-4 août 1914. Paris: Imprimerie Nationale, 1936.
Sometimes cited under Ministère des Affaires Etrangères.

France. Ministère de l'Instruction Publique et des Beauxarts. . . . Catalogue Général des Manuscripts des Bibliotheques Publiques de France. Vol. XLVIII, 2e Supplement, Rouen et Amiens. Paris: Plon, Nourrit & Cie, 1933.

Germany. Auswärtiges Amt. Die grosse Politik der europäischen Kabinette, 1871-1914. Berlin: Deutsche Verlagsgesellschaft für Politik und Geschichte, 1922-27. 40 vols.

Germany. (Federal Republic). Bundesministerium der Finanzen. Die Umsatzbesteurerung im Ausland. Finanzbericht, nr. 8, 10,11. T 1-3; Juni 1952-56. Bonn: Stollfuss [1957]. 3 vols.

Germany. Reichstag. Reichstags-handbuch, Vol. VII, 1932. Berlin: Reichsdruckerei, 1933.

9. PUBLICATIONS OF INTERNATIONAL ORGANIZATIONS

League of Nations.—The publications of the League of Nations include the records and documents concerned with the official activities and negotiations of the Council and Assembly; the factual, statistical, and descriptive materials issued by various Sections of the Secretariat in the fields of economics, political science, sociology, public health, etc.; the records and reports of related organizations and committees such as the Permanent Mandates Commission, or the International Institute of Intellectual Cooperation; and many of the reports of the autonomous and affiliated organizations such as the Permanent Court of International Justice, the International Labor Organization, and the International Hydrographic Office, the publications of which were issued also under imprints other than the League's. Publications of the League of Nations were printed in two official languages, French and English, which are found in parallel columns on the same page, on opposite pages, or in documents issued separately in each language. Some publications of the League continued to appear sporadically during the war, but on April 18, 1946, the League officially ceased to exist, and its publications, along with other assets, were transferred to the United Nations. Students, however, will continue to use and refer to material contained in League publications.

The *Official Journal* and its *Special Supplements* contain Council Minutes, Records and Resolutions of the Assembly, and most of the collateral documents. They are usually listed in the bibliography under League of Nations in order to keep all League references together. The bibliography may be arranged alphabetically or chronologically, after the corporate author. Specific citations of the Minutes, Records, etc., are made in the footnotes (see p. 92).
League of Nations. Official Journal. 20th Year, 1939.

The League of Nations publications were issued and distributed primarily to facilitate the execution of official business, and at first bore official numbers which indicated the order of their distribution to the different branches. Thus, C. 186.M.97.1937.VIA. reveals that the document was the 186th distributed to the Council and the 97th distributed to the Members in 1937. These numbers differ because a large number of documents were distributed to the Council which were not distributed among the Members uniformly. The document was distributed by Section VI A (Mandates). If a document originated in a League special committee, its initials and appropriate figures appear as part of the official number. In 1926, inauguration of the policy of assigning supplementary "publication sales numbers" to documents which were published separately and distributed generally (placed on sale) brought together by subject all publications of a given Section with consecutive numbering throughout each year. The sales number 1936.IX.16 identifies the 16th document to be distributed by Section IX in 1936. In bibliographical listing, these numbers are written in series position, usually with the roman numeral as the sole identification of the Section. It is recommended that, where the title of the document does not identify the Section, further identification be made in this manner: Disarmament.1936.IX.16. The official number was often retained (always on documents in the *Official Journal*) along with the sales number, but citation of the official number will rarely be necessary. If the official number were retained in the example just given, the whole number would read:

C.426.M.135.1936.IX.16. Usually, the official number may be disregarded; the sales number is sufficient to identify documents issued by the various Sections of the Secretariat. The sales number is often written: Ser. L. o. N. P.1936.IX.16., but if the League's auspices are shown by the corporate author, imprint, or otherwise, this expansion is not considered necessary. Some writers use the designation: League Document 1936.IX.16. If the League is shown elsewhere in the reference, it is not necessary to repeat League of Nations in the imprint. (Cf. "Alternate Form of Document Reference," p. 37.)

Illustrations

League of Nations. **National Control of the Manufacture of and Trade in Arms. Information as to Present Position Collected by the Secretariat in Accordance with the Resolution Adopted on May 31st, 1937, by the Bureau of the Conference.** Disarmament.1938.IX.1. Geneva: 1938. 241 pp.

Published also as Conf.D.184, i.e., Document 184 of the Disarmament Conference.

---- **Passport System. Replies from Governments to the Enquiry on the Application of the Recommendations of the Passport Conference of 1926.** Communications and Transit. 1938.VIII.3. Geneva: 1938. 17 pp.

League of Nations. Permanent Mandates Commission. **Minutes and Reports of the Thirty-third Session, November 8th-19th, 1937.** 1937.VIA.4. Geneva: 1937. 192 pp.

Special identification of Section VI A is not given, because the corporate author identifies it.

---- **Ratification of Agreements and Conventions Concluded under the Auspices of the League of Nations.** Nineteenth List. Legal.1938.V.4. Geneva: 1938. 139 pp.

This document is published also as: A.6(a).1938. Annex I. V., and as: *Official Journal, Special Supplement,* no. 181. Any or all of these identifications may be used, but the simple one used in the illustration is sufficient.

---- **The Recreational Cinema and the Young.** Social Questions.1938.IV.13. Geneva: 1938. 31 pp.

---- **Report of the Co-ordination Committee on the Economic and Financial Questions Contained in the Agenda of the Nineteenth Ordinary Session of the Assembly.** Economic Relations.1938.IIB.3. Geneva: 1938. 91 pp.

This document is published also as: A.16.1938.II.

<u>League of Nations. Statistical Yearbook,</u> 1942-44. Financial
and Economic Intelligence.1945.IIA.5. Geneva: 1945. 315
pp.

---- <u>Report on the Work of the League During the War. Submitted to the Assembly by the Acting Secretary-General.</u>
General.1945.2. Geneva: 1945. 167 pp.

    This Section has no assigned number. Often the series note is simply written: 1945:2.

Article 18 of the Covenant of the League of Nations provides that "every treaty or international engagement entered into...by any Member of the League shall be...registered with the Secretariat and shall as soon as possible be published by it." All such treaties and international engagements are published in the *Treaty Series* of the League of Nations. The *Annexes* contain information on ratifications, accessions, denunciations, etc. Bibliographical listing requires only the title (similar to statutes), but volume numbers may be given if desired. Specific citation is made in the footnotes (see p. 93). The *Treaty Series* is often cited without reference to the League of Nations, and consequently such references are often mistaken for the *Treaty Series* published by the U. S. Department of State. Therefore, it is suggested that when it is necessary to cite this *Treaty Series,* the words League of Nations be included arbitrarily as part of the title.

<u>Treaty Series,</u> vols. 201-3.

    Or, preferably:

<u>League of Nations Treaty Series,</u> Vols. CCI-CCIII.

League publications issued at fixed intervals, the articles in which are identified by volume, date, and page, take the form outlined for articles in periodicals (pp. 20 ff.).

Taylor, R.M., and others. "Investigation on Undulant Fever in France," <u>Bulletin of the Health Organization of the League of Nations,</u> VII (June, 1938), 503-45.

"Rations in Calories of 'Normal Consumers' in the Autumn of 1943," <u>League of Nations Bulletin of Statistics,</u> XXV (May, 1944), 129-31.

International Labor Organization.—The activities of many international agencies were expanded and coordinated with League of Nations enterprises under Articles 23 and 24 of the Covenant;

new agencies were established.[31] The various relationships that developed between these agencies and the League are reflected in documents issued by the agencies and by the League. The International Labor Organization was established in 1919 and became associated with the League as an autonomous organization. Its publications cover a wide range of subjects and are issued as series, yearbooks, periodicals, reports, etc. They are described and illustrated here as representative of the documents issued by international agencies. League of Nations documents concerned with the activities of these agencies are cited as League documents in the manner described above.

Although the activities of the International Labor Organization were coordinated with League interests, its publications were issued independently, and are treated separately in references. Membership in the League of Nations carried with it membership in the International Labor Organization. Provision was made also for participation by nonmembers, which allowed United States representation in the Governing Body and in the yearly International Labor Conference. The International Labor Office is the Secretariat of the Organization and publishes documents growing out of the activities of the Conference and Organization. Bibliographical references to these publications follow the principles for bibliographical references to government documents (see pp. 29 ff.). Entries are made under International Labor Office as corporate author. The spelling "Labour" appears in its publications, but in the United States "Labor" is used except when, in references, a title is reproduced exactly. The Series A to P of Studies and Reports on safety, social insurance, industrial hygiene, etc., should be identified in the series note, with appropriate numbers. During the war, the Office operated with headquarters in Montreal, which accounts for the variation in imprint in the illustrations given below.

---

[31] Hans Aufricht, *Guide to League of Nations Publications: A Bibliographical Survey of the Work of the League, 1920–1947* (New York: Columbia University Press, 1951), 682 pp. This work describes the status of affiliated, autonomous, and other related international agencies.

Illustrations

International Labor Conference. *The International Seamen's Code: Conventions and Recommendations Affecting Maritime Employment Adopted . . . 1920-36.* Montreal: 1942. 56 pp.
---- *Summary of Annual Reports under Article 22 of the Constitution of the International Labour Organization.* Geneva:: 1939. 872 pp.
---- ---- *Appendix. Report of the Committee of Experts on the Application of Conventions.* Geneva: 1939. 23 pp.
International Labor Office. *Economical Administration of Health Insurance Benefits.* Studies and Reports, Series M, no. 15. Geneva: 1938. 332 pp.
---- *The International Standardisation of Labour Statistics.* Studies and Reports, Series N, no. 25, Revision of no. 19. Montreal: International Labor Office, 1943. 169 pp.
---- *Labour Courts: An International Survey of the Judicial Systems for the Settlement of Disputes.* Studies and Reports, Series A, no. 40. Geneva: 1938. 220 pp.
---- *Minutes of the Ninety-first Session of the Governing Body,* London, December 16-20, 1943. [Montreal: International Labor Office, 1944.] 192 pp.
---- *Year-book of Labour Statistics, 1933/44.* Montreal: 1945. 265 pp.

Articles in periodicals published by the ILO are listed in the bibliography either by author or by title, as outlined on pages 20 ff.

"Collaboration of Employers and Workers with Government Departments in Great Britain," *International Labour Review,* LIV (Nov.-Dec. 1946), 321-30.
Tait, D.C. "Social Aspects of a Public Investment Policy," *International Labour Review,* XLIX (Jan. 1944), 1-18.

Permanent Court of International Justice.—Article 14 of the League of Nations Covenant provided for the establishment of a Permanent Court of International Justice "to hear and determine any dispute of an international character which the parties thereto submit to it." The Court (popularly called the World Court) was created by a separate Statute in 1921 and was set up at The Hague. The advisory opinions and judgments of the Court, the documents and proceedings concerning cases, the Court's annual report, etc., are contained in a series of publications (A/B to F in French and English), identification of which is made in a series note (cf. pp. 10–12). These documents were published by Sitjhoff in Leyden, and not by the League of Nations as might be expected. Specific page citations are made in the footnotes (p. 94).

Illustrations

Permanent Court of International Justice. <u>Consistency of Certain Danzig Legislative Decrees with the Constitution of the Free City.</u> Advisory Opinion of December 4th, 1935. Series A/B, no. 65. Leyden: Sitjhoff, [1936]. Pp. [39]-73.

> The Pleadings, Oral Statements, and Documents concerning this case are contained in Series C, no. 77, which may be included in the reference above or cited separately.

---- <u>Legal Status of Eastern Greenland.</u> Judgment of April 15, 1926. Series A/B, no. 53. Leyden: Sitjhoff, [1935]. Pp. [19]-147.

> Pleadings, Oral Statements, and Documents are contained in Series C, nos. 62-67.

---- Statute of the Court. <u>Rules of Court as Amended on July 31st, 1926.</u> Series D, no. 1. Leyden: Sitjhoff, 1926. 82 pp.
---- <u>Fifteenth Annual Report, June 15, 1938-June 15, 1939.</u> Series E, no. 15. Leyden: Sitjhoff, [1939]. 312 pp.

U. N. Conference on International Organization.—The United Nations Conference on International Organization which convened in San Francisco, April 25, 1945, represented the fulfillment of many preliminary negotiations which laid the groundwork for the establishment of an international organization to maintain peace and security. The documents which record the earlier steps (Moscow Declaration, Yalta Agreement, Dumbarton Oaks Proposals, etc) may be found in the documents of those countries which were parties to the negotiations. But the documents which originated in the Conference itself form a new category, with United Nations Conference on International Organization (place and date may be included) as the corporate author. The principal documents of the Conference have been published in sixteen volumes (English and French), and this compilation forms a standard source for citation.[32] The title of this collection, with an indication

---

[32] *Documents of the United Nations Conference on International Organization, San Francisco, 1945* (New York and London: United Nations Information Organization in Cooperation with Library of Congress, 1945–46).

of the volumes or committee reports consulted, will usually give sufficient information for listings in the bibliography; specific citation of individual documents will generally be found in footnotes (see pp. 94–95).

United Nations Conference on International Organization.
   Documents . . . Commission III, Security Council. London and New York: United Nations Information Organizations in Cooperation with the Library of Congress, 1945. Vols. 11-12.
---- Journal, nos 1-53, April 25-June 27, 1945. [San Francisco: 1945.]

U. N. Preparatory Commission.—Although the Charter was signed by fifty nations on June 26, 1945, the United Nations did not formally come into existence until October 24, 1945, when the requisite number of instruments of ratification had been deposited with the Government of the United States, and the Protocol of Ratification had been issued by the Secretary of State. However, in order that the preliminary work of preparing the various agenda and organizing the many organs of the United Nations might proceed, an agreement was signed, at the conclusion of the Conference, which established a Preparatory Commission and charged it with making the necessary provisional arrangements, including recommendations for the "transfer of certain functions, activities, and assets of the League of Nations."[33] The documents recording the activities of this Preparatory Commission form a second category of United Nations documents and should be identified accordingly. Until the United Nations was formally organized and the location of its headquarters agreed upon, the printing of documents was handled expediently. The original documents of the Preparatory Commission and those of the early meetings of the First Session of the General Assembly, Commissions, etc., were actually printed by His Majesty's Stationery Office, but this is not always shown in the imprint.

Bibliographical reference to the journals should include the number of the journal, date, and pagination; inclusion of the im-

---

[33] *Interim Agreements Concluded by the Governments Represented at United Nations Conference on International Organization,* Doc. 1190 G/127, June 26, 1945 (San Francisco: 1945), pp. 2–3.

print is optional. If individual documents are cited, the identifying symbol is given as a series notation. See the second example below. The following illustrations give a full form of bibliographical reference. Specific page citation is made in footnotes (p. 95).

Illustrations

United Nations. Preparatory Commission. Journal, nos. 1-27, 24 November-24 December 1945. London: [H. M. S. O. 1946.] 146 pp.
---- ---- Report of the Executive Committee of the Preparatory Commission of the United Nations. Document PC/EX/113/Rev.1., 12 November 1945. [London: 1945.] 144 pp.
---- ---- Summary Record of Meetings of Committee 2 - Security Council, nos. 1-10, 24 November-24 December 1945. London: [H.M. Stationery Office, 1946.] 30 pp.
---- ---- Summary Record of Meetings of Committee 6 - League of Nations, nos. 1-25, 24 November-24 December 1945. London: [H.M. Stationery Office, 1946.] 56 pp.

United Nations.—The General Assembly convened in London, January 10, 1946, to consider the *Report* of the Preparatory Commission and to initiate other business. At this First Session, six committees were created to assist the General Assembly in carrying out its work; the nonpermanent members of the Security Council were elected; the members of the Economic and Social Council were elected and in turn undertook the organization of commissions and negotiations for agreements with specialized agencies, as provided for in the Charter. Congress passed a formal resolution inviting the United Nations to establish its permanent home in the United States, and the invitation was accepted. In this period of organization, the scheme of symbols (documentation system) employed for the identification of documents was neither fully coordinated nor consistent. Of necessity, even the imprints varied with the location of the temporary headquarters—London, the Bronx, Lake Success, etc. While jurisdictions of the various bodies were being defined and their work patterns established, no scheme of symbols for the identification of documents could be final. Nevertheless, the principles of document citation applied in conformity with the discussion and illustrations in this section should serve as safe guides to the citation of United Nations documents at this early stage of development.

New York City was chosen as the permanent home, and the United Nations was installed in 1951 in the buildings which now comprise its headquarters. By this time, the organization had been achieved and was functioning, publications routines had been tested and regulated, and the documentation system had been stabilized. Inevitably, however, as the activities of the United Nations multiplied, it became necessary to add new symbols, and to modify or even to discontinue some of those in use. Such changes were generally made in conformity with the documentation system described here.

The United Nations is a vastly more complex organization than the government of any single country or of the League of Nations. The concerns of more than one hundred Member States and of the specialized intergovernmental agencies generate diverse activities which cover every continent and almost every human activity, and touch the lives of all of us in one way or another. The student should be aware that United Nations documents are produced to facilitate the participation of Members in the execution of official business, and that the complicated matters under consideration require elaborate and comprehensive information materials as well as *Official Records,* minutes, reports of deliberations, resolutions, etc. Each document deals with some phase of an issue or problem under consideration by one of the principal or subsidiary organs of the United Nations. It may represent either background information or reflect a step in the progression of an issue under consideration, from its appearance as a provisional agenda item through committee and plenary deliberations to its adoption as a resolution. Consequently, a knowledge of the structure, functions, and procedures of the United Nations and its component bodies is a primary requirement for a clear understanding of the significance of the documentation system, and of the identification and citation of documents of the General Assembly, the three Councils, and the bodies subordinate to them, such as main, standing, procedural, subsidiary, and *ad hoc* committees and commissions.[34]

---

[34] The Charter is readily available and should be familiar to anyone using United Nations documents. Leland M. Goodrich and Edvard Hambro, *Charter of the United Nations: Commentary and Documents* (2d and rev. ed.;

Some knowledge of the reference tools and guides, checklists of documents, bibliographies, and indexes to the documents of the various United Nations agencies will enable the student to make intelligent use of the vast amount of research material available in these publications and will aid in making unambiguous citation of the documents.[35]

United Nations documents are issued initially in mimeographed[36] preliminary (provisional) working versions for the use of Member States and of the Secretariat. Mimeographed documents are the primary United Nations source material: they are used by the U.N. delegates and constitute the most basic and comprehensive record of United Nations deliberations. A mimeographed document may be unique, that is, never reproduced in a final printed version. For these reasons, and because of the time lag between the provisional and final printed form of a document, they are sometimes indispensable to current or specialized research.

Much of this mass of mimeographed documentary materials is

Boston: World Peace Foundation, 1949), 710 pp.; and *A Bibliography of the Charter of the United Nations* (1955), 128 pp., U.N. Doc. ST/LIB/SER.B/3, provide background materials. *Everyman's United Nations* (7th ed.; New York: 1964), 638 pp., U.N. Sales No.; 64.I.9 is an excellent compact guide to the structure and functions of the United Nations and its related agencies. *Yearbook of the United Nations* (New York: 1947–), U.N. Sales No.: 64.I.1, is a comprehensive reference tool that serves as a continuing source of information on the structure and activities of the United Nations, the International Court of Justice, and related agencies. It includes organizational charts and references to the relevant U.N. documents. *United Nations Bulletin* (1946–1954), *United Nations Review* (1954–1964), and *UN Monthly Chronicle* (1964–) are official periodicals which will be useful for supplementary and current proceedings. A bibliography of secondary works is found in Amry Vandenbosch and Willard N. Hogan, *Toward World Order* (New York: McGraw-Hill, [c.1963]), pp. 319–28.

[35] A wide range of such publications is issued by the Secretariat and other agencies. The use of these and more specialized tools is described in Brenda Brimmer, and others, *A Guide to the Use of United Nations Documents*, N.Y.U. Libraries Occasional Paper, no. 3 (Dobbs Ferry, N.Y.: Oceana Publications, 1962), Pt. II. This work will be indispensable to the advanced student and research worker.

[36] As newer duplicating processes have been developed, the term "mimeographed" has been applied to documents reproduced by means other than printing.

corrected, edited, and republished in final printed versions, in the *Official Records* (Supplements or Annexes) of the General Assembly or of the three Councils (see pp. 68–69), or as United Nations sales publications.[37] (see pp. 64–68). The final printed versions are generally more available and should be cited in references unless they are inadequate for research needs, or there is some cogent reason for citing preliminary documents.

The United Nations documentation system provides for the complete identification of each individual document.[38] This system is based upon a sequence of symbols which reflects the organizational structure of the United Nations and the procedures for conducting its business—meetings, conferences, deliberations, etc., of the main organs and subsidiary bodies. The symbol scheme combines capital letters and arabic numerals. A diagonal stroke separates the components of the symbol. Basic, secondary, and descriptive symbols make up the whole symbol.

Documents are usually issued in the working languages, English and French (Spanish in the General Assembly and the Economic and Social Council). The final printed versions are printed in the official languages—English, French, Spanish, Russian, and Chinese. (Some publications are available in Arabic and Portuguese.) Each mimeographed document is assigned a symbol and a date which are the same for every language edition. This symbol is retained when a document is republished in the *Official Records* (or as a sales publication). Facility in the use and citation of U.N. documents requires recognition of the symbols employed and an understanding of the significance of their location in the sequence of symbols.

In accordance with Article 7 of the United Nations Charter six principal organs were established. The documents of these principal organs were given characteristic basic symbols: A/ for the

---

[37] The terms United Nations documents and United Nations publications are used interchangeably in this work.

[38] *List of United Nations Document Series Symbols*, ST/LIB/SER..B/5/Rev.1 (1965), 139 pp., Sales No.: 65.I.6; *United Nations Documents Index*, ST/LIB/SER.E/CUM.3, Vol. I–   ), issued monthly; Harry N. M. Winton, "Documentation," *Annual Review of United Nations Affairs*, 1949 (New York: N.Y.U. Press, 1950), pp. 52–68; Brimmer, and others, *op. cit.*, pp. 5–19.

General Assembly, S/ for the Security Council, E/ for the Economic and Social Council, and T/ for the Trusteeship Council. The documents of the two remaining principal organs, the Secretariat and the International Court of Justice, and their identifying symbols are described separately below on pages 70–75, and documents of specialized agencies on pages 75–77.

The basic symbols above identify the primary source of a document. A numeral added after a diagonal slash identifies the individual document. For example, A/71 identifies the 71st document of the General Assembly; E/15 identifies the 15th document issued by the Economic and Social Council.

It should be kept in mind that United Nations is still a developing institution and that it must continually adapt to changing needs. The documentation system allows for expansion as United Nations activities increase, but it has not always been possible to make entirely consistent additions to this scheme. The Charter did not provide for some special bodies that were later created in response to new demands. The Atomic Energy Commission (AEC/) was created by the General Assembly in January 1946 because of the sudden importance of the problems of atomic energy. It was dissolved in 1952 when the General Assembly created the Disarmament Commission (DC/) to carry on its work (along with that of the Commission for Conventional Armaments). In 1959, the General Assembly and the Economic and Social Council created the United Nations Special Fund (SF/) which has developed a comprehensive technical assistance program throughout the world. Other special bodies include the United Nations Administrative Tribunal (AT/) and the Military Staff Committee (MSC/). Although these special bodies are not principal organs, their documents are identified by basic symbols. For example, DC/12 identifies the 12th document issued by the Disarmament Commission.

Secondary symbols supplement the basic symbols of the principal (parent) organs and special bodies, and identify a ramification of subordinate committees, sub-committees, *ad hoc* committees, commissions, etc. A secondary symbol is made up of capital letter(s) and an arabic number, and follows the basic symbol after a diagonal slash: A/C.4 designates the 4th Committee of the General

Assembly; A/C.4/6 identifies the 6th document issued by the Committee. Subordinate bodies are customarily numbered in the order of their establishment. More than one secondary symbol may be used. For example, E/CN.1/SC.2/12 identifies the 12th document issued by the 2d sub-committee of the 1st Commission of the Economic and Social Council. The basic symbol must always be the first component of the whole symbol, as is shown in the following illustrations.

```
Main, Standing, or Permanent Committees... DC/C.1
Sub-committees ......................... S/C.1/SC.1
Ad hoc committees....................... A/AC.82
Conferences ............................ A/CONF.30
Preparatory committees.................. E/CONF.5/PC.3
Commissions ............................ A/CN.4
Sub-commissions ........................ E/CN.6/SUB.2
Working group........................... E/CN.12/C.1/WG.1
Working party .......................... E/CONF.37/C.2/WP.2
```

As United Nations activities have multiplied, it has been necessary to establish additional committees, commissions, etc. The practice of assigning conventional secondary symbols has not, however, been strictly followed. Consequently, inconsistencies have developed in the sub-series of symbols just described. In some cases the initials of the descriptive names (acronyms) of the subsidiary bodies replace the conventional capital letter(s) and number(s) symbol; in other cases an acronym is used as an alternative symbol. It is necessary to use the symbol A/SPC to identify the Special Political Committee of the General Assembly because a conventional secondary symbol has not been assigned to this committee. Similarly, the Economic Commission for Europe can only be identified /ECE, but the Economic Commission for Africa may be alternatively identified by the conventional symbol /CN.14, or by the acronym /ECA. In either case the symbol sequence remains the same and is always secondary to the basic symbol of the principal organ. Such variations should not cause confusion since it is reasonable to assume sufficient familiarity with the body whose documents are being cited so that its initials are easily recognized. For example, the whole symbol E/TAC/AC.2/31 should readily translate: 31st document of the 2d *ad hoc* committee of the Technical Assistance Committee, a subsidiary body of the Economic and Social Council.

A group of descriptive symbols is common to all principal organs and special and subsidiary bodies. This sub-series (sometimes called additive) designates the character of the documents. Representative descriptive symbols are shown below in examples using combinations of basic and secondary symbols.

Add.	(Addendum)	SF/18/Add.1
Agenda	(Agenda)	DC/Agenda/53
Corr.	(Corrigendum)	E/C.3/PC.2/Corr.1
CR.	(Communications)	A/CN.7/CR.5
INF.	(Information series)	S/INF./5
L.	(Limited distribution)	E/C.2/L.25
Min.	(Minutes)	A/C.6/Min.1
NGO	(Non-governmental organization)	E/CN.5/NGO/36
PET.	(Petition)	T/PET./5
PV.	(Verbatim records, i.e., Procés-verbaux)	A/C.5/PV./51
Res.	(Resolution)	A/Res./822 (IX)[39]
Rev.	(Revision)	A/C.2/L.782/Rev.1
SR.	(Summary records)	T/SR./702

U.N. Sales Publications.—Following the practice of the League of Nations (pp. 51–53), a selected number of United Nations printed publications are made available for sale by authorized agents throughout the world. These publications do not include the various *Official Records, Treaty Series,* or periodicals (e.g., *The United Nations Review*). Each sales publication is assigned a sales number which is printed in a box on the back of the front cover or of the title page.

The whole sales number consists of three parts: the year of publication, the category in roman numerals, and the number of the particular publication in arabic numerals, e.g., 65.XV.9[40] identifies the 9th publication in the category Human Rights issued in 1965. Seventeen broad subject categories have been established[41] and are represented by the roman numerals in the sales number, e.g.,

---

[39] The roman numeral indicates the number of the session.

[40] Since 1958 it has been the practice to omit the century.

[41] *United Nations Publications, 1945–1963,* ST/CS/SER.J/3, (1964), 70 pp. and subsequent sales catalogs list publications in categories; *United Nations Document Index,* see footnote 38, includes sales publications.

II. identifies the category, Economics and Finance. Further subdivision of this category is identified by capital letters: II.A (Economic Stability and Employment), II.B. (Economic Development). The third publication issued in 1967 in the subcategory II.H. (Technical Assistance) would be identified by the sales number, 67.II.H.3.

The sales number provides a convenient identification of sales publications intended for the general public. Inevitably, however, some sales publications will carry the conventional United Nations series symbols in addition to the sales number. This double identification occurs when the sales publication has been used previously as a mimeographed working document by some U.N. body. Confusion is avoided in bibliographical references by placing the sales number at the end of the reference (if it is necessary to include it). One encounters a variety of citations of sales numbers: U.N. Sales No., U.N.P. Sales No., United Nations Publication, Sales No., etc. In a bibliography where United Nations is part of the reference, it is not considered necessary to repeat the U.N. identification.

Bibliographical references to United Nations documents should be made in conformity with the principles of citation discussed earlier for government documents (pp. 29 ff.). Because general acceptance of international standards of bibliography and documentation has not been realized, a great variety of bibliographical citations of United Nations documents will be found. The bibliographical items recommended here are necessary for identification of a document, are nonrepetitive, and will enable the reader to locate the document in a library card catalog. Strict adherence to this sequence of bibliographical detail is, however, not necessary. Sometimes a different sequence is used to serve a specific purpose.[42] If a different sequence is adopted, it should be used consistently, and redundant information should be avoided.

---

[42] *Bibliographical Style Manual,* U.N. Doc.: ST/LIB/SER.B/8 (1963), 62 pp., Sales no.: 63.I.5, is a lucid treatment of standards of bibliographical style for the guidance of the Secretariat. United Nations documents are discussed and illustrated on pp. 35–50. Please note that this citation is clarified by the insertion of "U.N. Doc." because United Nations does not occur otherwise.

The complexity of the Organization of United Nations and its many cooperative relationships may present difficulty in determining the corporate author. The proper form of entry will be found in the *United Nations Document Index*,[43] the *National Union Catalog*,[44] or the library card catalog. Personal authors are practically never evident in United Nations publications. If there should be occasion to cite a document by personal author, the form for government documents on pages 30–31 should be used. Long titles may be shortened intelligently by the use of three dots(. . .).

The United Nations document symbol follows the title. This symbol is found identified as: U.N. Doc. or Doc., etc. Such designations are necessary only if the United Nations aspect has not been shown in other parts of the reference. This need usually occurs in footnotes (see footnote 42, and page 96). The symbol is sometimes within parentheses, but this distinction is not considered to be necessary.

New York, as the place of publication is ordinarily omitted. If, however, the place of publication is other than New York,[45] it must be designated (see the third and seventh illustrations below, and specialized agencies, pp. 76). The publisher is generally considered to be unnecessary in the imprint because the corporate author will have identified the publisher and repetition of United Nations does not serve any purpose. The date of publication must be included, if it differs from the date in the title of the document.[46] Total or inclusive pagination is often omitted in references, but here it is considered desirable to allow the reader to know the length of the reference (see pp. 14–15).

If the sales number is given, it is the last item in the reference. It may be identified as: Sales No., or U.N. Sales No., if necessary. It is

---

[43] See footnote 38.

[44] (Washington: Library of Congress, 1956–   ).

[45] The four Regional Commissions sometimes issue documents with a Geneva (ECE), Addis Ababa (ECA), Santiago, Chile (ECLA), or Bangkok (ECAFE) imprint.

[46] In some specialized works the registration date is required, that is, the date on which the document was registered by United Nations Documents Control.

an identification for sales purposes and many authors omit it in book and article references.

Illustrations

United Nations. Disarmament Commission. Report of the Conference of the Eighteen-Nation Committee on Disarmament, 21 Jan. - 17 Sept. 1964. DC/209. 1964. 57 pp. E/F

E/F indicates that this document is bilingual, i.e., printed with English and French arranged in parallel columns. Consequently, the English text amounts to about 28 pages, not 57 pp. as shown in the reference. Cf. citation of this document in the *Official Records*, page 69.

---- Economic and Social Council. Commission on Human Rights. The Question of an International Code of Police Ethics. E/CN.4/NGO/117. 1964. 5 pp.

---- Economic Commission for Asia and the Far East. Bauxite Ore Resources and Aluminum Industry of Asia and the Far East. Mineral Resources Development Series, no. 17. E/CN.11/598. Bangkok: 1962. 51 pp. 64.II.F.2.

---- Economic Commission for Latin America. La Industria Textil en America Latina: Chile. E/CN.12/622. 1963. 97 pp. 63.II.G.5.

---- General Assembly. Administrative and Budgetary Coordination between the UN and the WHO . . . A/3596. 1957. 49 pp.

---- ---- Question of Southern Rhodesia. Working Paper Prepared by the Secretariat 9 October 1964. A/C.109/L.150. 19 pp. incl. Annexes.

The date of publication is omitted because it is the same as the date in the reference.

---- ---- 6th Committee. Plans for the Formulation of the Principles Recognized in the Charter of the Nuremberg Tribunal and in the Judgment of the Tribunal . . . A/C.6/180. Lake Success: 1946. 8 pp.

---- Secretariat. The Status of Permanent Sovereignty over Natural Wealth and Resources Study. A/AC.97/5/Rev.2; E/3511 (A/AC.97/13). 1962. 245 pp. 62.V.6.

It is not unusual for a document to carry the symbols of two organs, if the business under consideration is the concern of both, or for other reasons.

---- Secretary General. Economic and Social Consequences of Disarmament. Report of the Secretary General. E/3593/Rev.1; E/3593/Rev.1/Add.1-5. 1962. 304 pp. 62.IX.1-2.

---- Security Council. The Palestine Question. S/PV449. 1961. 14 pp. E/F

---- ---- Committee of Experts. Draft Rules of Procedure of the Security Council Relating to Voting Proposed by the

68   Bibliography and Footnotes

 <u>Delegation of the United States.</u> S/C.1/160. Lake Success: 1948. 22 pp.
 ---- Trusteeship Council. <u>Report of the UN Visiting Mission to the Trust Territories of Togoland under British Administration and . . . . French Administration.</u> T/1238. 1956. 44 pp., map.

Official Records are published separately for the General Assembly, each of the three Councils, the Atomic Energy Commission, and the Disarmament Commission. They comprise verbatim or summary records of plenary or committee meetings, and are found on library shelves bound into sets of volumes for each session.[47] References should be listed in the bibliography under United Nations, followed by the name of the organ. The words *Official Records* are italicized because this is the set to which reference is made. General references that lack any indication of the pertinent parts of the *Official Records* are not acceptable. Page citations should be given when only a portion of a volume is used. Plenary and committee meetings should be indicated when necessary. Session number, meeting number(s), date(s), and the total or inclusive pages cited are given in the forms shown in the illustrations. The dates should be given in the same form used on the title page of the document. Abbreviations should be used only to clarify references. Cf. footnote citation, page 97.

Illustrations

United Nations. Atomic Energy Commission. <u>Official Records.</u> 4th Year. Nos. 3-4, 19th-20th Meetings, 15-22 March 1949. 17, 16 pp.
---- Economic and Social Council. <u>Official Records,</u> 28th Session. 1085th-1089th Meetings, 24-31 July 1959. pp. 133-164.
---- General Assembly. <u>Official Records,</u> 16th Session. Plenary Meetings, Vol. II. 1067th Meeting, 28 November 1961. pp. 882-890.
---- ---- <u>Official Records,</u> 15th Session. Second Committee. Summary Records, 694th Meeting, 3 December 1960. pp. 385-93.

---

[47] Since 1950, each meeting record is published separately in fascicle form (i.e., a small number of printed sheets bound together and issued at intervals). The fascicles are paged continuously so that later they may be easily incorporated into a bound volume with the identical paging.

Bibliography   69

---- Security Council. <u>Official</u> <u>Records,</u> 16th Year. 949th
Meeting, 11 April 1961. 14 pp.

Security Council *Official Records* are identified by year (not session number), because this body functions continuously.[48] Each meeting record is separately paged.

---- Trusteeship Council. <u>Official</u> <u>Records,</u> 12th Session.
461st-473rd Meetings, 16 June-7 July 1953. pp. 1-185.

In a long list of United Nations publications it may be desirable to differentiate the references by identifying the subject.

United Nations. Security Council. <u>Official</u> <u>Records,</u> 15th
Year. 857th-863rd Meetings, 23 May 1960. "U-2 Incident."
Various pagings.
---- ---- ---- 880th Meetings, 22-26 July 1960. "RB-47
Incident." Various pagings.

The *Supplements* and *Annexes* form a valuable adjunct to the various *Official Records*. They contain individual reports of main and subsidiary U.N. bodies, resolutions, decisions, and selected documents essential to an understanding of matters under consideration. Bibliographical references include the title of the individual document (within quotation marks), followed by the U.N. symbol and the *Official Records* identification. Abbreviations are used in the illustrations below to avoid ambiguity, and also to avoid repeating the name of the organ.

Illustrations

United Nations. Disarmament Commission. "Report of the Conference of the Eighteen-Nation Committee on Disarmament,
21 Jan-17 Sept 1964." DC/209. 57 pp. (DC <u>Official</u> <u>Records,</u>
<u>Supplement</u> for Jan. to Dec., 1964.) E/F
---- Economic and Social Council. "Non-Governmental Organizations." E/3261. 1959. 1 p. (ESC <u>Official</u> <u>Records,</u> 28th
Session. <u>Annexes</u>: Agenda Item 16).
---- General Assembly. "Admission of the Republic of Cyprus
to Membership." Resolution 1489 (XV), 20 September 1960.

Annexes are compiled consecutively by Agenda Item numbers which must be given in references. The date has been given in this reference because it is not shown otherwise.

---

[48] *United Nations Official Records, 1948-1962: A Reference Catalogue,* ST/CS/SER.J/2 (1963), 107 pp., 64.I.3, is a comprehensive list of the *Official Records* of all U.N. organs, with explanations of their publication practices.

A/4462. 63 pp. (GA *Official* *Records*, 15th Session, *Supplement*, no. 11.).
---- ---- "Annual Report of the Secretary General on the Work of the Organization 15 June 1962-15 June 1963," and "Introduction to Annual Report . . . 20 August 1963." A/5501; A/5501/Add.1. 150, 7 pp. (GA *Official* *Records*, 18th Session, *Supplement*, no. 1 and 1A.).
---- ---- "Question of Race Conflict in South West Africa Resulting from Policies of Apartheid . . ." A/4804/Add. 1-5. 1961. 23 pp. (GA *Official* *Records*, 16th Session. *Annexes*, vol. 2: Agenda Item 76.).
---- Trusteeship Council. "United Nations Visiting Mission to Trust Territories in East Africa, 1957; Report on Somaliland under Italian Administration." T/1404. 38 pp. (TC *Official* *Records*, 22d Session, *Supplement*, no. 2.).

U.N. Secretariat.—Pursuant to Article 7 of the Charter, the Secretariat was established as a principal organ of the United Nations. The Secretary General, chief administrative officer of United Nations, is also the executive head of the Secretariat. Its increasing importance corresponds to the ever expanding activities of United Nations. The fundamental difference of function between the Secretariat and the other principal organs demands documents of a different type and a different document identification.

The General Assembly and the three Councils were established as deliberative, policy and decision making bodies composed of the delegates of Member States. The delegates are national representatives who are responsible to their own governments. On the other hand, the Secretariat is primarily an administrative, coordinating, and service agency with a staff of non-partisan international civil servants, who are responsible only to the Organization of the United Nations. The staff serves the other United Nations organs; it collects information for them, administers and coordinates their programs, implements policies laid down by them, manages their meetings, prepares background materials for the meetings, handles "housekeeping" and financial matters, etc.[49] Documents produced by the Secretariat staff to meet these needs often appear under the

---

[49] Vandenbosch and Hogan, *op. cit.*, pp. 82–93. See also the organization chart in Sydney D. Bailey, *The Secretariat of the United Nations*, United Nations Study, no. 11 (New York: Carnegie Endowment for International Peace, 1962), pp. [74–75].

various document symbols of the other organs. For example, *The Role of Patents in the Transfer of Technology to Underdeveloped Countries* (1961) is a report of the Secretary General prepared in the Secretariat, but this document carries the symbols, E/3861; E/C.5/52Rev.1. Documents concerned with the internal administration of the Secretariat or with budget and financial affairs of the Organization generally bear the symbol of the General Assembly: *Budget Estimates for the Financial year 1964 and Information Annexes,* submitted by the Secretary General, A/5505.

Documents are issued by the Secretariat in its own name. The sub-series of secondary symbols which identify by number the various committees and commissions of the other principal organs do not provide appropriate identification for these Secretariat documents. Each department and office of the Secretariat is identified by its own individual secondary symbol which is made up of its initials or fairly obvious abbreviations, e.g., ST/SG (Executive Office of the Secretary General), ST/LEG (Office of Legal Affairs, ST/LIB (United Nations Library, i.e., Dag Hammarskjold Library), ST/OPI (Office of Public Information, since 1958, previously ST/DPI was used for Department of Public Information). The various series of specialized and general bibliographies, indexes, and other aids to United Nations research, and publications produced for the use of the Secretariat are identified in the same way.[50]

Descriptive symbols identify the various lettered series of documents issued by each office and department of the Secretariat, e.g., ST/LEG/SER.A, ST/LIB/SER.F. Administrative instructions and information circulars issued by the Secretary General are identified /AI and /IC. The descriptive symbols /Rev. (revision), and /Add. (addendum) are used when appropriate.

Press releases have their own particular system of symbols. Although they are issued by the Secretariat (Office of Public Information, Bureau of Press and Publications), they are generally considered to be outside of the scope of United Nations documents and are not included here.

---

[50] *List of United Nations Document Series Symbols,* see footnote 38, lists all the symbols in use.

Bibliographical citation of Secretariat documents should be made in accordance with the recommendations for United Nations documents (pp. 65–66). In addition to the illustrations below, footnotes and earlier illustrations in this section show various examples of Secretariat documents.

Illustrations

```
United Nations. Dept. of Economic and Social Affairs. New
    Sources of Energy and Economic Development: Solar Energy,
    Wind Energy, Tidal Energy, Geothermic Energy and Thermal
    Energy of the Seas. E/2997; ST/ECA/47. 1957. 150 pp. 1957.
    II.B.1.
---- ---- Statistical Office. Classification of Commodi-
    ties by Industrial Origin. Statistical Papers, Series M,
    no. 43. ST/STAT/SER.M/43. 1966. 72 pp.
---- Office of Legal Affairs. Legislative Texts and Treaty
    Provisions Concerning the Utilization of International
    Rivers for other Purposes than Navigation. ST/LEG/SER.
    B/12. 1963. 934 pp. 63.V.4.
---- Secretary General. Annual Report of the Secretary Gen-
    eral on the Work of the Organization 15 June 1962-15 June
    1963. A/5501. 150 pp.
```

Cf. citation of this document in the *Supplement* to the *Official Records* of the General Assembly, page 70.

```
---- ---- Introduction to Annual Report . . . 20 August
    1963. A/5501/ADD. 1. 7 pp.
---- ---- Historical Survey of the Question of Interna-
    tional Criminal Jurisdiction: Memorandum submitted by the
    Secretary General. A/CN.4/7/Rev.1. Lake Success: 1949.
    147 pp. 1949.V.8.
---- Secretariat. United Nations Juridical Yearbook, 1963.
    ST/SER.C/1. 263 pp. 65.V.3.
```

Article 2, section 1 of the Charter prescribes that: "Every treaty and every international agreement entered into by any Member of the United Nations ... shall as soon as possible be registered with the Secretariat and published by it." Continuing the practice of the League of Nations (see p. 53) all treaties, conventions, agreements, annexes, exchanges of notes, etc. are published in the *United Nations Treaty Series* after they have been deposited with and registered by the Secretariat's Office of Legal Affairs. The full title should be given to differentiate this publication from other treaty series. In bibliographical citation only the title and volume(s) need to be given. Specific citation is made in the footnotes (see page 98).

United Nations Treaty Series, vols. 396, 403.

International Court of Justice.—The International Court of Justice is the principal judicial organ of the United Nations; it is seated at The Hague. The Court provides the means of resolving problems and settling legal disputes among states (especially Member nations). Its jurisdiction includes all cases which parties refer to it and all matters specifically provided for in the Charter, treaties, or conventions in force. Advisory opinions are given by the Court on all matters referred to it by the General Assembly, the Security Council, and other authorized bodies.

The Court functions in accordance with a Statute that is an integral part of the Charter, but it is an autonomous body in certain respects, e.g., it issues and distributes its own publications which are printed in Leyden by A. S. Sitjhoff's Publishing Co. The publications are organized in several categories which continue the series pattern used for the publications of the Permanent Court of Justice (cf. pp. 55–56). (1) Decisions given by the Court are published in *Reports of Judgments, Advisory Opinions, Orders.* (2) The documentation and collateral material for each case is contained in *Pleadings, Oral Arguments, Documents.* (3) The Charter, Statute of the Court, Rules of Court, etc. are published in *Acts and Documents Concerning the Organization of the Court.* (4) An annual account of the work of the Court with summaries of judgments and advisory opinions given during the year is found in the *Yearbook.*[51] (5) A bibliography of critical and interpretative works relating to the Court is compiled annually in the *Bibliography* (issued as Chapter IX of the *Yearbook* until 1963–1964).

Serial sales numbers which are assigned consecutively provide complete identification of individual volumes and simplify citation.[52] Usage varies, but it is recommended that bibliographical

---

[51] *ICJ Yearbook,* No. 19, 1964–65, pp. 141–142, Sales No.: 296, describes the Court's publications.

[52] Cf. *A Uniform System of Citation* (10th ed.; Cambridge: Harvard Law Review Association, 1958), pp. 74–75, sec. 21:1.

citation include the title of the category (italicized), the date, volume number (when necessary), short title of the case,[53] and sales number. Imprint is generally considered to be unnecessary, but may be included for clarity. If any purpose is served, the contending parties may be indicated as shown in the first illustration below. Specific citation is made in the footnotes (see p. 98).

```
International Court of Justice. Reports, 1964. "South West
   Africa." (Ethiopia v. South Africa; Liberia v. South Af-
   rica.) Sales No.: 289.
---- Pleadings, 1950, Vol. I. "Certain Norwegian Loans."
   Sales No.: 189.
---- Yearbook, Nos. 18-19, 1963-1965 The Hague: 1965. Sales
   Nos.: 278, 296.
```

The International Court of Justice should not be confused with the International Law Commission which was established by the General Assembly in 1947 to promote the development and codification of international law. Commission sessions are held in Geneva. Its *Yearbook* (published by United Nations) contains meeting records (Vol. I), and selected documents including the *Report* of the Commission to the General Assembly (Vol. II). The *Yearbook* is listed in the bibliography as a United Nations document; the Report may be cited as part of the *Yearbook*, or as a General Assembly document.

```
United Nations. International Law Commission. Report . . .
   Covering Work of Its 15th Session 6 May-12 July, 1963.
   A/5509.
---- Yearbook, 1966. 2 vols. A/CN.4/SER.A/1966. Sales No.:
   66.V.1.
```

The periodicals issued by the various bodies of the United Nations cover almost all phases of its work. Bibliographical references take the same form as articles in any other periodical (see p. 20). The author's name is written in inverted order; the title of the periodical is italicized (underlined); volume, date, and inclusive paging indicate the location of the article in the periodical and its length. Reference to specific pages is made in the footnotes (see p. 98).

---

[53] *ICJ Yearbook*, No. 19, 1964–65, pp. 132–134 lists the short titles adopted by the Court.

"Index Numbers of Value of Retail Trade." **Monthly Bulletin of Statistics** (U.N.), XX(September, 1966), pp. 92-95.

U.N. has been inserted in order to differentiate this publication from other "bulletins of statistics."

Milam, Carl H. "Work of the Library Committee." **United Nations Bulletin**, V(September 1, 1948), 681-82.

Specialized agencies.—Article 57 of the Charter provides that: "The various specialized agencies...having wide international responsibilities, as defined in their basic instruments, in economic, social, cultural, educational, health, and related fields shall be brought into relationship with the United Nations..." Specialized agencies are autonomous organizations related to the United Nations by individual intergovernmental agreements which are negotiated with the Economic and Social Council. The General Assembly may be concerned with budgetary matters, may suggest subjects for study or investigation, and approves final agreements.

Relationships vary in accordance with individual agreements, but do not involve the surrender of powers. Unlike United Nations subsidiary organs, each specialized agency is an independent legal entity directed by its own governing body, and has its own membership, administrative organization, budget, programs, and publications.[54] Specialized agencies working with each other and with United Nations perform a wide variety of functions. This intricate system of international cooperation is maintained and coordinated by the Economic and Social Council. Agreements between the United Nations and more than a dozen specialized agencies are presently in force.

The names of most of the specialized agencies are familiar: International Labor Organization (ILO), United Nations Educational, Scientific, and Cultural Organization (UNESCO), Food and Agricultural Organization (FAO), World Health Organization (WHO), International Civil Aviation Organization (ICAO), etc. Their relation to the United Nations is not, however, widely

---

[54] *Everyman's United Nations*, Pt. III describes the organization, functions, membership, and activities of specialized agencies. *Yearbook of the United Nations*, Pt. II, contains an annual account of activities, and publications are included in the *United Nations Document Index*.

understood and this causes abstruse references to be made to their publications.

Documents concerning the establishment, functions, and activities of specialized intergovernmental agencies, or defining their relationship with United Nations or with each other may appear in any or all three of the following categories: (1) As documents of the individual specialized agencies, (2) as documents of member governments, or (3) as United Nations documents.[55] Bibliographical references to documents in categories (1) and (2) are made in the form recommended for government documents (see pp. 29 ff.); those in category (3) in the form recommended for United Nations documents (see pp. 65 ff.).

The schemes of document identification vary with the practice of each agency. Annual reports, studies, documents issued in numbered series, etc., are cited in references in the forms recommended for government or United Nations documents, whichever is appropriate. The place of publication should be specified because these agencies maintain their headquarters in various parts of the world. The official forms of the names of the agencies will be found in the references cited in footnotes 34, 35, and 38, above.

Illustrations

Canada. Dept. of External Affairs. <u>International Telecommunication Convention and Related Documents, Signed at Atlantic City, October 2, 1947</u>. Treaty Series (1947), no. 33. Ottawa: 1948. 101 pp.
Food and Agricultural Organization of the United Nations. <u>The International Effects of National Grain Policies</u>. FAO Commodity Policy Studies, no. 8. Rome: 1955. 20 pp.
International Labour Conference, 46th Session, 1962. <u>Prohibition of the Sale, Hire, and Use of Inadequately Guarded Machinery</u>. Report VI (1). Geneva: 1961. 38 pp.
International Labour Office. <u>Indigenous Peoples, Living and Working Conditions of Aboriginal Populations in Independent Countries</u>. Studies and Reports, n.s., no. 35. Geneva: 1953. 628 pp.
International Monetary Fund. <u>Balance of Payments Manual</u>. Final Draft. 3d ed. Washington: 1960. 249 pp.

---

[55] Brimmer, and others, *op. cit.*, pp. 19–49, "The System of the Specialized Agencies"; pp. 184–232, "Tools and Guides to the Specialized Agencies."

International Telecommunication Union. List of Inter-Continental Telephone Communication Channels (Direct Circuits). ITU Publication, no. 73. 2d ed. Geneva: 1962. 19 pp. E/F/S

United Nations. Bureau of Social Affairs. International Survey of Programmes of Social Development. Prepared in Co-operation with the International Labour Office . . . E/CN.5/332; ST/SOA/39. 1959. 190 pp.

---- Economic and Social Council. Committee on Programme Appraisals. Five-Year Perspective, 1960-1964: Consolidated Report on the Appraisals of the Scope, Trend, and Costs of the Programmes of United Nations, ILO, FAO, UNESCO, WHO, WMO, IAEA in the Economic, Social and Human Rights Fields. E/3347/Rev.1. 1960. 120 pp. 60.IV.14.

---- Economic Commission for Asia and the Far East. Marketing Edible Oils (Liquid) and Oil-Seeds in the ECAFE Region. Prepared by the ECAFE/FAO Agricultural Division. E/CN.11/419. 1956. 47 pp. 1956.II.F.5.

---- General Assembly. Convention on the Privileges and Immunities of the Specialized Agencies . . . ST/LEG/4. 1953. 58 pp. 1953.X.1.

---- Secretary General. Agreements between the United Nations and the Specialized Agencies. ST/SG/1. 1952. 132 pp. 1951.X.1. E/F

United States. Dept. of State. Draft Charter for the International Trade Organization of the United Nations. Commercial Policy Series, no. 16. Washington: G.P.O., 1947. 87 pp.

Universal Postal Union. Emission de Timbres-Poste - Demande du Japon. Circulaire 180/1959. Berne: 1961. 4 pp.

World Health Organization. Decisions of the UN, Specialized Agencies and IAEA Affecting WHO's Activities on Financial Questions . . . 22 Feb 1961. WHO Resolution 14.29. Geneva: 1961. 1 p.

Most of the specialized agencies publish periodicals, bulletins, news letters, etc. References to articles are cited in the form recommended on pages 20-24.

"Aedes Aegypti Eradication in the Americas," WHO Chronicle, 18 (January, 1964), 3-8.

Maheu, René. "Universal Literacy: Time for Action," Statement made . . . to 2d Committee of the U.N. General Assembly 11 November 1965, UNESCO Chronicle, XI (Dec. 1965), 461-467.

Samuilenko, F., "Stabilization and Training of Manpower in the Forestry Industry of Byelorussia," International Labour Review, 83 (June, 1961), 523-546.

Regional organizations.—Regional organizations are a comparatively recent development in international organization. The Charter recognizes regional agencies (provided that their activities are consistent with the principles and purposes of the United Nations) and encourages their participation in the settlement of local disputes and in the maintenance of international peace and security. When the Charter was drafted the need for such activities at the regional level did not, however, appear to be significant because it was believed that once the United Nations was established and fully operative regional alliances would become unnecessary; that disputes would be settled and international peace would be secured at the universal level. But, in fact, the most striking development in the international system since 1945 has been the proliferation of regional organizations.[56] Military factors have fostered the growth of regionalism. Scientific and technological advances have generated a steadily increasing interdependence among the countries of the world, while new means of communication and transportation facilitate larger units of administration. A variety of regional arrangements has been devised to meet the rapidly changing needs, and the areas of operation have expanded to include not only military, but also economic, political, social, cultural, and nuclear research purposes.

The Charter does not provide a framework for the resulting network of supranational activities, nor define relationships between regional groups and the United Nations. Coordinating links are lacking, and there is a need to integrate regional functions and structures with the United Nations system. Inevitably, documents that are a by-product of this indeterminate situation will present problems in uniform citation.

Regional organizations are independent legal entities with membership made up of sovereign states. The term "regional" is used loosely and does not necessarily correspond with a recognized geographical area. (The United States belongs to regional organizations whose other members are located on all continents except Africa.) They show a striking diversity of institutions, powers, and

---

[56] Ronald J. Yalem, *Regionalism and World Order* (Washington: Public Affairs Press, [c.1965], p. 1.

procedures. Regional organizations are rooted in agreements of varying elaborateness widely scattered through publications that in some instances are difficult to obtain.[57]

This diversity of functions, interests, and activities of regional groups produces a wide range of publications in many languages—reports, studies, bulletins, news letters, series, treaties, etc. These documents may be issued by any member of the organization, or by the organization, and sometimes by the United Nations, or by any combination of these agencies in cooperation.

Bibliographical references to these documents should be made in the forms recommended for government documents (p. 29) or for United Nations documents (p. 65). The official names of the organizations (corporate authors) will be found in the *Yearbook of International Organizations*.[58] The place of publication should always be given in references, and if the publisher is other than the organization it should be given. Articles in periodicals published by regional organizations are cited in the same manner as articles in any other periodical (see pp. 20–24).

Illustrations

Australia. Proceedings of the South Seas Commission Conference, Canberra, 28th January - 6th February, 1947. Canberra:1947. 90 pp.
Benelux Economic Union. Étude Comparative des Budgets de l'État des Pays du Benelux. [Bruxelles]: 1961. 59 pp. E/D
---- ---- Supplement: Tableaux. 1961. 36 pp. E/D
Council of Europe. Administrative Arrangements for the Health Control of the Seas and Air Traffic Adopted in 1956 under the Auspices of Western European Union . . . Strassbourg: 1961. 7 pp. E/F
European Coal and Steel Community. High Authority. Der Schutz der Arbeitnehmer bei Verlust des Arbeitsplatzes, von G. Boldt, et al. Sammlung des Arbeitsrechts, 2. Das Arbeitsrecht in der Gemeinschaft, 11. Baden Baden: A. Lutzmeyer, [1961]. 529 pp.
European Economic Community. Commission. Proposals for the Working-out and Putting into Effect of the Common Agri-

---

[57] Ruth C. Lawson, ed., *International Regional Organizations: Constitutional Foundations* (New York: Praeger, 1962), pp. v–vi.

[58] (Brussels: Union of International Associations, 1948–   ) Pt. II, "The European Community"; Pt. III, "Other Inter-governmental Organizations."

cultural Policy in the Application of Article 43 of the Treaty . . . VI/Com(60)105. Brussels: 1960. Various pagings.

European Organization for Nuclear Research. Symposium du CERN sur les Accélérateurs de Haute Énergie et la Physique des Mesons π . . . 11-25 Juin 1956. Compte Rendus . . . CERN 56-25-26. Geneva: 1956. 2 vols.

Goslinga, W.J. "Netherlands Educational System Used in Caribbean," Caribbean Commission Monthly Bulletin, II (October, 1948), 67-68.

League of Arab States. Cultural Dept. A General Review of the Cultural Activities of the League . . . 1946-1956. Cairo: C. Tsoumas & Co. Press, [1957?]. 51 pp.

Organization for African Unity. Council of Ministers. Resolutions and Recommendations of the First Session . . . Held at Dakar, Senegal from 2 to 11 August 1963: Official Text. n.p.: 1963. 2 l.

Organization for European Economic Cooperation. Joint Trade and Intra-European Payments Committee. Liberalisation of European Dollar Trade: Second Report . . . Paris: [1957]. 180 pp.

Southeast Asia Treaty Organization. Council of Ministers. 11th Meeting, June 27-29, 1966. "Council Communiqué, June 29th," U.S. Dept of State Bulletin, LV (Aug. 1, 1966), 172-74.

U.S. Dept of State. Atomic Energy: Cooperation for Peaceful Uses. Agreement between United States . . . and the European Atomic Energy Community (EURATOM) Amending Agreement Signed at Brussels and Washington, May 21 and 22, 1962. Treaties and Other Acts Series, no. 5103. Washington: G.P.O., 1962. 36 pp.

United Nations. Economic Commission for Latin America. Foreign Private Investment in the Latin American Free Trade Area. Report of Consultant Group Jointly Appointed by the Economic Commission for Latin America and the Organization of American States. E/CN.12/550. 1961. 30 pp. 60.II.5.

FOOTNOTES

THE ENTRIES of the formal bibliography, at the end of a book, chapter, or article, which make reference to whole works, are differentiated from footnotes, which are specific references to a particular page, pages, chapter, or chapters. Each quotation, statement, fact, or idea taken from another author should be acknowledged in a footnote citation of the publication from which it was taken. Footnotes are used (*a*) to cite sources of information, (*b*) to give additional information concerning matters treated in the text, (*c*) to direct attention to supporting, divergent, or conflicting opinions, and (*d*) to refer to other pages or passages in the text (cf. pp. 1–3). In the absence of a formal biliography, the footnotes are the only guide to the supporting material used in the preparation of the work, and great care should be taken to note accurately the location of the information cited.

The explanatory type of footnote which amplifies the text should be used sparingly (see footnote 15, p. 16, which should be incorporated in the text; and footnote 22, p. 26, which gives more information than is pertinent). If the material is important enough to be noted at all, it should probably be incorporated in the text in proper continuity, where it will not distract the reader. If one feels compelled to elaborate a point, the use of notes at the end of the chapter, or further amplification in an appendix, is preferable to prolonged footnote interruptions.

Although this manual places particular emphasis on the footnote which cites specific pages or passages, it is recognized that occasionally it will be necessary to direct the reader's attention to an entire work or compilation. The type of footnote that serves this purpose occurs mainly in works intended for purposes of instruction. In footnote 23 on page 30 above, an entire compilation is cited for the reader's convenience. It is not intended that the reader should think the compilation was used as supporting data or source material; hence a specific citation would be inappropriate.

The documentation of a work bears evidence of the author's indebtedness to others, of his discrimination and judgment in the collection and use of supporting data, and of his willingness to subject his sources to investigation. Footnotes should be used judiciously. The presence of numerous footnotes is not necessarily an indication of the good quality of a work, but the author's selection, evaluation, and interpretation of the supporting data, and its pertinence as cited, do serve as evidences of quality, and ought to show good quality.

Honesty demands scrupulous care in the assignment of credit for borrowed materials; paraphrased, reworded, or rearranged statements do not evade this responsibility. Footnotes are the most important vehicle for conveying credit, but a précis or summary of the material, or a direct quotation, may be used and credited effectively within the text itself. Direct quotations should be accurate, closely related to the immediate text, and no longer than is necessary to illustrate the issue under discussion.

Footnotes are numbered consecutively throughout an article, chapter, or monograph of moderate length. The consecutive numbering of footnotes throughout an entire book should be avoided. Most books are formally organized by chapters or other well-defined subdivisions, and the footnotes should be numbered consecutively from "1" throughout each chapter or subdivision.

An arabic superior number (e.g., [1]) in the text refers the reader to a footnote which is headed by a corresponding superior number at the bottom of the page or further on. The superior number in the text should follow just after the word, statement, or quotation to which the footnote pertains. It should be placed outside of punctuation or quotation marks (e.g.,... the fundamental problem.[1] "...in great detail."[2]... according to most authorities,[3] [but one must take account of the exceptions][4]...).

Superior numerals are used in the text only. In order to avoid confusion in footnote references to tables, charts, graphs, etc., superior (superscript) letters of the alphabet are used, in alphabetical order from "a." Although the use of symbols should be limited, it is permissible to use asterisks, daggers, etc., with footnotes which refer to headings or subheadings of the text, or to distinguish one category of notes from another, e.g., the editor's notes from the author's. If it is necessary to use symbols, they should occur in the

1. BOOKS

The first footnote citation to a book should give complete bibliographical information, but specific reference should be made to the relevant passages or pages only (references to whole works can seldom be justified). The imprint is sometimes omitted in footnotes, or is shortened to two items, the place and date of publication. The place is relatively unimportant unless the name of the publisher accompanies it, but in many fields the date is essential. It is recommended that the imprint be given in full in the first footnote citation, as illustrated below. Full footnote citations should be in the following form (cf. bibliographical forms, pp. 16 ff. above):

[1]Author's Name in Regular Order, Title of Book in italics, Supplementary note, if necessary, Series and number, if any (Edition; Place: Publisher, Date), volumes in roman (capital) numerals, pages cited in arabic numerals, illustrations, etc., if necessary.

Illustrations

[1]Norman Angell and others, Economic Principles and Problems, ed. by W. E. Spahr (4th ed.; New York: Farrar and Rinehart, 1941), II, 129-31.
[2]Edward Kasner and James Newman, Mathematics and the Imagination, illustrated by Rufus Isaacs (New York: Simon and Schuster, 1943), p. 281.
[3]Frank MacShane, Many Golden Ages: Ruins, Temples . . . (Tokyo and Rutland, Vt.: Tuttle, [1963]). p. 23, pls., 8-9.
[4]Henry L. Mencken, The American Language . . . (4th ed., corr. and enl.; New York: Knopf, 1938), Supplement II (1948), pp. 385-6.
[5]Paul Nitze, U.S. Foreign Policy, 1944-1945, Headline Series, no. 116. ([New York: Foreign Policy Association], 1956), p. 47.

The volume-and-page citation illustrated in the first reference above is simple, yet does not permit ambiguity, and it eliminates the need of the abbreviations vol. and pp. Such a reference, however, will be found in varying forms, e.g.:

[1]Norman Angell and others, Economic Principles and Problems, ed. by W. E. Spahr, Vol. II (4th ed.; New York: Farrar and Rinehart, 1941), pp. 129-31.

This sequence could not be used in the fourth illustration because the Supplement must be shown as Supplement II of the fourth edition, and consequently, if notation of the Supplement does not follow a complete reference to the fourth edition, an explanatory note is required. The form recommended is preferable because it can be used, without change, to meet either situation.

In footnotes, citation of the volume and page or pages is specific, brief, and sufficiently informative. Additional designation by chapter or section is not necessary. References can become needlessly complicated if the volume and page citation are duplicated by other notations; for example: Vol. III, chap. 7, secs. 9–11, pp. 129–31, may be replaced by the simple notation III, 129–31, which serves the same purpose and is preferred. References to chapters, books, and sections might well be limited to citations for which custom or format makes them necessary, as for example the classics, law, or the large city newspaper. In the newspaper citation, section identification saves time in locating the article. When complicated references are justifiable or necessary, care should be taken to keep the references as compact as is consistent with the information to be conveyed.

The corporate author is not customarily given as the first item in a footnote citation (cf. bibliographical references, p. 17); but the institution or society responsible for the publication is shown in the title, series, imprint, or otherwise.

[1]<u>New Spanish Painting and Sculpture: Rafael Canogar,</u> Exhibition, Museum of Modern Art, (Garden City: Doubleday, [1960]), pp. 19-20.

[2]<u>Theatre Guild Anthology</u> (New York: Random House, [c. 1936]), pp. 560-78.

2. ARTICLES

The first footnote reference to an article should contain the same information as that given in a biliographical reference, but should be written with the modifications enumerated on pages 83–84. Entire articles are often cited, but the general reference can scarcely convey the precise information desirable; more often than not, it reflects lazy habits of mind and is not pertinent. Specific references to selected pages are preferable.

Full footnote citations to articles should be in the following form (compare with the form on pp. 22 and 23 above):

¹Author's Name in regular order, "Title of Article, in quotation marks," Name of Periodical in italics, volume of periodical in roman (capital) or arabic numerals as printed on the periodical itself (Date of issue in parentheses), page or pages cited.

Illustrations

¹Charles S. Holmes, "James Thurber and the Art of Fantasy," Yale Review, LV (Autumn, 1956), 29.
²Celeste Budd Horne, "A Geographer Looks at Russia," Current History, n.s., 8 (Feb., 1945), 147.
³David McCord Wright, "The Future of Keynesian Economics," American Economic Review, XXXV (June, 1945), 284-5.

The examples just given illustrate the forms recommended here. In this sequence the date separates the volume and page numbers—a separation which, in the opinion of many, produces a clearer reference than other arrangements do. The inclusion of the month in the date is optional. The illustration below shows an acceptable but not recommended variation (see p. 24).

¹Charles S. Holmes, "James Thurber and the Art of Fantasy," Yale Review (Autumn, 1956), 55: 29.

It is permissible to write the name of the periodical in abbreviated form when no ambiguity will result. Even skeleton abbreviations may be used in works of interest to a small and well-defined audience. *S.R.* (*Saturday Review*) or PMLA [*Publications of the Modern Language Association*] would be intelligible to students of literature, but these abbreviations should be amplified in a work intended for general interest, for example, *Sat. Rev.* and [*Pub. Mod. Lang. Assoc.*] Standard abbreviations found in the various periodical indexes should be used.

3. NEWSPAPER ARTICLES

The first footnote citation of newspaper articles includes the same items as are given in bibliographical references (see pp. 24 ff. above). Although page references are commonly omitted, it is recommended that they be included.

Illustrations

¹San Francisco Chronicle, Jan. 23, 1964, p. 7.
²"Nationalizing the Coal Industry" (editorial), Times (London), Dec. 21, 1945, p. 4.
³Fiorello La Guardia, "Why New York Should Be the World Capital," P M (New York), Feb. 3, 1946, p. 3.
⁴Aline B. Saarinen, "The National Gallery: Chester Dale, President and Financier Collector . . ." New York Times, May 6, 1956, sec. 6, pp. 14, 33.
⁵"Terrible Railroad Disaster," New York Times, Jan. 23, 1865, p. 8, col. 2.

4. PARTS OF BOOKS—ENCYCLOPEDIA ARTICLES

The information given in bibliographical references to parts of books or to encyclopedia articles (pp. 26 ff.) should be included in full footnote citation, but the forms should be adapted to the modifications enumerated on pages 83–84.

Illustrations

¹Charles C. Colby, "The Role of Shipping in the World Order," in Walter H.C. Laves, ed., The Foundations of a More Stable World Order, Harris Foundation Lectures, 1940 (Chicago: University Press, 1941), pp. 92-3.
²Felix Frankfurter, "Benjamin Cardozo," Dictionary of American Biography, Supplement Two, XXXII (1959), p. 194.
³"Brook Farm," Columbia Encyclopedia, 3d ed. (1963), p. 279.
⁴Bergen Evans, "Tale of a Tub," in his Natural History of Nonsense (New York: Knopf, 1946), pp. 258-75.

In citations like the second and third just above, the word "in" is sometimes inserted before the name of the encyclopedia. Although the word "in" eliminates the possibility of confusion of authors' names in the Colby illustration, and "in his" avoids repetition of the author's name under Evans, no advantage would be gained by the addition of "in" to the other references.

5. INDIRECT QUOTATIONS

The source of a quotation should be accurately acknowledged, and the original is always to be preferred to a secondary authority. If one is obliged to use a quotation, not from the original work, but as found quoted by another author, this should be shown in a footnote citation as follows:

```
¹Gamaliel Bradford, The Journals of Gamaliel Bradford,
1883-1932, ed. by Van Wyck Brooks (Boston: Houghton Mif-
flin, 1933), p. 207, as quoted by Monroe E. Deutsch in The
Letter and the Spirit (Berkeley: University of California
Press, 1943), p. 134.
```

6. DISSERTATIONS, MANUSCRIPTS, AND INTERVIEWS

Footnote citations to theses or dissertations follow the form recommended for books, except that the thesis statement and the date are enclosed in parentheses. The total number of leaves is usually given in the bibliographical reference, but the specific page citation in the footnote may be given with the abbreviations p. or pp.

```
¹Charles W. Shumaker, English Autobiography: Its Materi-
als, Structure, and Technique (Ph.D. dissertation, Univer-
sity of California, 1943), pp. 112, 189.
```

Footnote references to manuscript materials are written in paragraph form and should include information enough to identify and locate the manuscript. Citation should be by author, if the authorship is known. No part of the citation mentioned in the text need be repeated in the footnote. The peculiarities of manuscript materials are described on pages 19–20.

```
¹Edwin Arlington Robinson, "The Man Against the Sky,"
Snyder Collection, Williams College Library, Williamstown,
Mass., second MS leaf gives line arrangement as indicated.
²"Amadís de Gaula," Lansdowne MSS 766, fols. 1-18b, Brit-
ish Museum, London, Eng.
```

References to interviews or personal letters are made in the footnotes, and should include only those details which are necessary to supplement the information in the text. Often the footnote may be written simply:

```
¹Information obtained in correspondence with Mr. Ackerley.
²Statement made to the author by Dr. Taylor.
```

The date, circumstances of the interview, or other details are given when they have any significant importance for the text discussion.

References to unpublished letters usually omit the word "letter," but they commonly include the abbreviation L.S. (letter, signed) or A.L.S. (autograph letter, signed), if either applies. These abbreviations are sometimes found as Ls and ALs or l.s. and a.l.s. The note tells also where the letters are to be found.

¹H[enry]. R. Wagner, Berkeley, Calif., to [Charles N.] Kessler, [Helena, Mont.], Mar. 23 [1919?], A.L.S., 2 pp., The Montana Papers, no. 165, William Andrews Clark Memorial Library, Los Angeles, California.

7. U. S.—NATIONAL, STATE, and MUNICIPAL DOCUMENTS

The corporate author is commonly omitted in footnote citation of government documents. Reference is made by title; the notation of the series and the facts of publication are so arranged as to indicate the official source of the document cited; and the footnote is written in paragraph form with proper indention. In the first illustration given below, U. S. Bureau of Foreign and Domestic Commerce has been included in the series notation. It is not necessary, however, to amplify the third and seventh illustrations, because in each the official source of the document is shown in the title. These footnote illustrations should be compared with the bibliographical references illustrated on pages 34–37, 40.

A uniform system of footnote citation to government documents cannot provide adequately for the wide diversity of materials and the variations of government publication practices. These complexities are more easily handled by the use of the corporate author in bibliographical references. In the customary absence of the corporate author in footnote citation, care must be taken to include information enough to identify the source of the document. Occasionally, clear identification will require the inclusion of the corporate author. The circumstances and good judgment, rather than rigid adherence to the forms illustrated below, should indicate the items to be included. For example, on page 30 above, the titles of the government publications are given in the text and are not repeated in the footnotes. However, to insure proper identification of the documents cited, the corporate author is included in footnotes 24 and 25. Omission of the corporate author carries little risk of confusion when a work contains both footnotes and a formal bibliography; but if the bibliography is lacking, or if it contains only selected references, the risk increases and should be offset by including the corporate author in the first footnote citation of a document.

Illustrations

¹<u>Small Retail Store Mortality,</u> by William T. Hicks and Walter F. Crowder, U.S. Bureau of Foreign and Domestic Commerce, Economic Series, no. 22 (Washington: G.P.O., 1943), pp. 17-19.

²<u>Economic Report of the President . . . to Congress January 1966 . . .</u> 89th Cong., 2d sess., H. Doc. 348, (Washington: G.P.O., 1966), p. 253.

³<u>Annual Report of the Inland Waterways Corporation to the Secretary of Commerce, Calendar Year 1943</u> (Washington: G.P.O., 1944), pp. 5-8.

⁴<u>. . . Report on Unemployment Insurance to the Fifty-sixth California Legislature,</u> Senate interim Committee on Unemployment (Sacramento: 1945), p. 132.

⁵<u>Chicago Zoning Ordinance Passed by the City Council . . . May 29, 1957, as Amended Nov. 25, 1958</u> (Chicago: Index Pub. Corp., 1959), p. 217B.

⁶<u>Atomic Energy,</u> Hearings on S. 1717, U.S. House Military Affairs Committee, 79th Cong., 2d sess. (Washington: G.P.O., 1946), pp. 7-9.

⁷<u>Message of the Mayor of the City of New York to the Board of Estimate . . . Executive Budget for the Fiscal Year, 1941-42</u> (New York: 1942), p. 7.

⁸<u>Message of Governor William G. Stratton to the 71st Assembly, January 7, 1959,</u> ([Springfield, Ill.: 1959]), p. 19.

The first of these illustrations may be cited by the authors (cf. Wright and Macmahon, pp. 30–31). In references to government documents, main words need not be written with capital letters (e.g., ...*Report on unemployment insurance to the fifty-sixth*... or, *Chicago zoning ordinance passed by the city council*...).

8. BRITISH GOVERNMENT DOCUMENTS

The rules for footnote citation given above for United States documents apply here. The examples below should be compared with the bibliographical references on pages 45–47.

Illustrations

¹"Cardinal Aquaviva to James III," <u>Calendar of the Stuart Papers . . .</u> Hist. MSS Comm. (London: H.M. Stationery Office, 1921), VII, 347-8.

²<u>Report by H.M. Government . . . to the Council of the League of Nations on the Administration of Palestine and Trans-Jordan, 1938.</u> Colonial, no. 166 (London: H.M.S.O., 1939), p. 346.

³Ernest Satow, <u>International Congresses.</u> Foreign Office Peace Handbooks, Vol. XXIII, no. 151 (London: H.M.S.O., 1920), pp. 68, 137.

⁴<u>Agreements . . . Regarding Financial Assistance to Czecho-Slovakia . . . Jan. 27, 1939.</u> Parl. Pubs., 1938-39, Vol. XXVII (Accounts and Papers, vol. 12), Cmd. 5933 (London: H.M.S.O., 1940), p. 11.

    May be cited also as *Treaty Series,* no. 9(1939).

⁵<u>Post-war Policy Private Woodlands. Supplementary Report of the Forestry Commission.</u> Parl. Pubs., 1943-44, Vol. III (London: H.M.S.O., 1945), p. 5.

    This footnote is given in the simplest form compatible with clarity. Commissioners' Reports, vol. 1, if added, would facilitate location of the reference on library shelves.

⁶<u>A Bill to Prohibit the Hunting with Hounds of Deer; to Provide for the Control of Deer by Approved Methods . . .</u> Parl. Pubs., 1959-60, Vol. III, H.C.B. 125 (London: H.M.S.O., 1960), p. 2.

⁷<u>A Bill Intituled an Act to Consolidate with Amendments Certain Enactments Relating to Local Government in London.</u> Parl. Pubs., 1938-39, H.L. Sess. Papers, Vol. III, H.L. 103 (London: H.M.S.O., 1939) p. 167.

    Either of the bills named may be cited by short title—*Land for Planning Purposes,* or *Local Government in London*—and the citation may be to sections of the bills in place of the page numbers.

⁸<u>Wheat Fund Accounts, 1942-43.</u> Parl. Pubs., 1943-44, Vol. V, H.C. 47 (London: H.M.S.O., 1944), p. 3.

    Accounts and Papers, vol. 1, would, if added, make a more nearly complete reference.

9. INTERNATIONAL ORGANIZATIONS

League of Nations.—The corporate author is customarily omitted in footnote references to League documents. Sufficient identification of the corporate author should, however, be given in the series note, imprint, or otherwise. References to the *Official Journal* are treated in the same manner as references to articles in periodicals. The examples below should be compared with the full forms illustrated on pages 52–53.

Illustrations

¹"Minutes of the 103rd Session of the Council, September 26th, 1938," League of Nations Official Journal, 19th Year (Jan. 6, 1938), 867.
²"Texts Adopted by the Inter-American Conference for the Maintenance of Peace, Buenos Aires, December 1st-23rd, 1936," Official Journal, Special Supplement, no. 178 (Geneva: League of Nations, 1937), pp. 53-4.
³Passport System. Replies from Governments to the Enquiry on the Application of the Recommendations of the Passport Conference of 1926. L. of N. Doc., Communications and Transit.1938. VIII.3 (Geneva: 1938), p. 4.
⁴Minutes and Reports of the Thirty-third Session, Nov. 8th, 1937 [Permanent Mandates Commission], L. of N. Doc. 1937.VIA.4 (Geneva: 1937), pp. 187-8.
⁵Statistical Yearbook of the League of Nations, 1942-44, L. of N. Doc. Economic.1945.IIA.5. (Geneva: 1945), p. 256.

The *Treaty Series* references give the volume number, page, and registry number (cf. p. 53). The volume number is usually given in roman numerals in order to differentiate it from the page numbers. A full form of reference is seldom necessary, but the last illustration below is given with complete identification of the document.

¹ 9 Treaty Series, 415, Reg. no. 269.

Or:

¹League of Nations Treaty Series, IX, 415, Reg. no. 269.

Or:

¹Treaty Series, 415, Reg. no. 269.
²"Exchange of Notes Regarding the Grant of Naval and Air Facilities to the Government of the United States of America in British Transatlantic Territories, and the Transfer of United States Destroyers to the Government of the United Kingdom, Washington, September 2d 1940," League of Nations Treaty Series, CCIII, 201-7, Reg. no. 4762.

Articles in periodicals published by the League or related organizations are cited in footnote references in the same manner as other periodical articles, by author or by title as is necessary (see p. 53).

¹R.M. Taylor and others, "Investigation on Undulant Fever in France," Bulletin of the Health Organization of the League of Nations, VII (June, 1938), pp. 519-20.

²"Rations in Calories of 'Normal Consumers' in the Autumn of 1943," *League of Nations Bulletin of Statistics*, XXV (May, 1944), 130.

International Labor Organization.—Footnote references are treated in the same manner as U. S., League, or other documents described above (cf. pp. 52–54).

Illustrations

¹*The International Seamen's Code: Conventions and Recommendations Affecting Maritime Employment Adopted by the International Labor Conference, 1920-36* (Montreal: 1942), p. 5.
²*Economical Administration of Health Insurance Benefits*, I.L.O. Studies and Reports, Series M. no. 15 (Geneva: 1938), pp. 153-61.
³*Minutes of the Ninety-first Session of the Governing Body, London, December 16-20*, 1943 [Montreal: International Labor Office, 1944], p. 119.
⁴*Year-book of Labour Statistics, 1943/44* (Montreal: International Labor Office, 1945), pp. 214-15.
⁵D.C. Tait, "Social Aspects of a Public Investment Policy," *International Labour Review*, XLIX (Jan. 1944), 9-10.

Permanent Court of International Justice.—Footnote references to advisory opinions seldom give more than the name of the Court, series, identifying number within the series, and the specific pages cited.

¹Permanent Court of International Justice, *Series A/B*, no. 65, p. 29.

If fuller references are desirable, any part or all of the full form on pages 55–56 may be adapted to footnotes. In the short citation above, the series is italicized, but in a full reference which includes the title of the judgment, advisory opinion, etc., the title is italicized.

¹*Fifteenth Annual Report of the Permanent Court of International Justice, June 15, 1938 - June 15, 1939*. Series E, no. 15, (Leyden, Sitjhoff, [1939]), pp. 279-80.

U. N. Conference on International Organization.—Most of the Conference documents originated in the four Commissions (I-IV) which were created to develop general principles and recommend proposed texts for adoption as parts of the Charter. Technical committees (1–4) were created within the Commissions to prepare draft provisions of parts of the text of the Charter. Thus, the document symbol "II/4/3" identifies Commission II, committee 4, and the third document issued by this committee. Various initials were used to identify other classes of documents and these also were combined with numbers. G was used for general material, DC for reports of delegation chairmen, EX for executive committee, WD for working document, etc. The symbols identify individual documents, and, in the reference, are placed in series position. It is not necessary to give the imprint, but the bibliography should always indicate the work or collection which includes the document, e.g., the sixteen-volume compilation referred to on page 56, or *The United Nations Conference on International Organization...Selected Documents*,[2] etc.

Illustrations

[1]*Meeting of the Heads of Delegations to Organize the Conference, April 27, 1945,* UNCIO Doc. 30, DC/5 (1), p. 15.

[2]*Report of Mr. Paul Boncour, Rapporteur, on Chapter VIII, Section B,* UNCIO Doc. 881, III/3/46, p. 9.

[3]*Documentation for Meetings of Committee IV/2: Relation of International Law and the Charter to Internal Law,* UNCIO WD no. 12, IV/2/24, p. 2.

U. N. Preparatory Commission.—These documents are cited in the same manner as U.N.C.I.O. documents. In the absence of corporate author, identification of the Commission should be given elsewhere in the reference.

[1]*Journal of the United Nations Preparatory Commission,* no. 11, Dec. 6, 1945 ([London: H.M.S.O.], 1946), pp. 54-5.

[2]*Report of the Executive Committee to the Preparatory Committee of the United Nations,* PC/EX/113/Rev. 1 (London: [H.M.S.O.], 1945), p. 134.

United Nations.—Students are advised to use the form of footnote reference to United Nations documents which is illustrated below, unless special circumstances dictate otherwise. In specialized works, the reader of which is expected to be familiar with

---

[2] U. S. Dept. of State, Conference Series, no. 83 (Washington: G.P.O., 1946), 991 pp.

these source materials, the citation is sometimes reduced to document symbol or sales publication number. For general purposes, however, a more nearly complete identification is required.

Footnote references to United Nations documents customarily omit the corporate author. If the official source of the document is not evident in the title, the series notation or facts of publication may be arranged to identify the issuing agency. When the corporate author, e.g., the Economic and Social Council, is mentioned in the part of the text to which the footnote applies, it should not be repeated in the footnote citation. Standard abbreviations may be used. The footnotes are written in paragraph form and give specific page citation. The sequence of items recommended for bibliographical citation of United Nations documents adapts easily to the form for footnote citation. The illustrations below should be compared with the bibliographical references on page 65 ff., and with the footnotes which cite U.N. documents.

Illustrations

[1] Report of the Conference of the Eighteen-Nation Committee. U.N. Doc.: DC/209 (1964), p. 33.
   U.N.: is added before the document symbol for clarification.

[2] Plans for the Formulation of the Principles Recognized in the Charter of the Nuremberg Tribunal and in the Judgment of the Tribunal . . . U.N. Doc.: A/C.6/180 (Lake Success: 1947), p. 2.
   The 6th Committee of the General Assembly is identified by the symbol; it may be further identified as Legal Committee, if desirable.

[3] Bauxite Ore Resources and Aluminum Industry of Asia and the Far East. Mineral Development Series, no. 17, U.N. Doc.: E/CN.11/598 (Bangkok: 1962), Sales No.: 64.II.F.2. p. 17.
   Economic Commission for Asia and the Far East (ECAFE) is identified by CN.11 in the symbol, but the name of the Commission or its initials may be added before the series note, if it serves any purpose.

[4] Administrative and Budgetary Co-ordination between the UN and WHO . . . U.N. Doc.: A/3596 (1957), p. 32.

[5] The Status of Permanent Sovereignty over Natural Wealth and Resources Study. U.N. Doc.: A/AC.97/5/Rev.2; E/3511 (A/AC. 97/13) (1962), Sales No.: 62.V.6.
   Many authors omit the sales number in references. Others would expand it by adding, U.N Pub., and might even include the subject covered by the

category number (roman numeral, see pp. 64–65): U.N.Pub., Sales No.: International Law, 64.V.6.

The Official Records of the main organs of United Nations are printed with varying designations on the title page. To avoid making long and involved references, titles are shortened by using three dots (...). Abbreviations are used to keep references compact. They should not, however, be used indiscriminately, and should be used consistently throughout a work. For example, if the abbreviation "GA *Official Records*" is adopted, it should not be shortened a few pages later to "*GAOR*". The examples below give a full form of footnote citation which is not always necessary. Citations in specialized works may safely be made in a shorter form. Books and articles intended for general use should give enough particulars for easy identification of a document. *Official Records*, *Supplements*, and *Annexes* are italicized; titles of reports, resolutions, etc., are enclosed within quotation marks. Parentheses set off the title from the citation to *Official Records*. If the title is omitted, the parentheses become unnecessary. The session (or year), meeting date(s), and specific page citation are given. It will sometimes be necessary to indicate that verbatim or summary records are being cited. Cf. bibliographical references, page 68 ff.

Illustrations

[6]"Question of China in the United Nations" (UNGA Official Records, 16th Session, Plenary Meetings, Vol. II, 1071st Meeting, 9 December 1961), p. 923.

In specialized works U.N. may be omitted, or might even be shortened: GAOR.: 16 Sess., V.II, 1071st Pl. Mtg., 9 Dec. 1961. This abbreviated form of footnote is, however, unsatisfactory for most readers.

[7]"Diplomatic Intercourse and Immunities," Note by the Secretary General, 28 July 1959 (UNGA Official Records, 14th Session, Annexes: Agenda Item 56), p. 23.

[8]"Investigation of Alleged Bacterial Warfare" UNSC Official Records, 7th Year, 588th Meeting, 8 July 1952), p. 15.

[9]"Report of the Eighth Session of the Commission on International Commodity Trade, 2-13 May, 1960" (UNECS. Official Records, 30th Session, Supplement, no. 6), p. 5.

[10]"Admission of the Republic of Cyprus to Membership," Resolution 1489 (XV), 20 September 1960, A/4462 (UNGA Official Records, 15th Session, Supplement, no. 11), par. 3.

Footnote citations to the *United Nations Treaty Series* should include the volume number, page, and registry number (cf. pp. 72–73. Occasionally it is necessary to include the specific title in a fuller form, which is illustrated in the last of the three examples following.

```
¹U.N Treaty Series, V, 206, Reg. no. 30.
```

Or:

```
¹United Nations Treaty Series, V, 206, Reg. no. 30.
²"Convention between the United States of America and the
United States of Mexico for the Solution of the Problem of
the Chamizal (and Annexes, and Exchange of Notes), Signed
at Mexico on 29 August 1963." United Nations Treaty Series,
vol. 505 (1965), 217 and map. E/S
```

Articles in the periodicals issued by subordinate bodies of the United Nations are cited in footnotes in the form recommended for periodical articles on pages 86–87.

```
¹"Index Numbers of Value of Retail Trade," Monthly Bulle-
tin of Statistics (U.N.), XX (September, 1966), 94.
²Carl H. Milam, "Work of the Library Committee," United
Nations Bulletin, V (September 1, 1948), 681.
```

The International Court of Justice publications are generally cited in an abbreviated form characteristic of legal citation. ICJ is added before the reference to identify the Court. A shortened form of the full series title, i.e., *ICJ Reports of Judgments, Advisory Opinions, Orders,* is given with the date, volume number (if necessary), and page citation.

```
¹ICJ Reports, 1964, p. 18.
```

The fuller form of footnote citation illustrated below is often desirable in general works, and can be adapted to suit specific purposes. For explanations of the short title, sales number, etc., see pages 73–74.

```
¹"South West Africa," Order of 20 October 1964, ICJ Re-
ports, 1964, Sales No. 289, p. 171.
²"South West Africa," Preliminary Objections, Judgment,
ICJ Reports, 1962, Sales No. 270, p. 319.
³"Certain Norwegian Loans," ICJ Pleadings, 1950, Vol. I,
Sales No. 189, pp. 64-65.
⁴ICJ Yearbook, No. 19, 1964-1965, Sales No. 296, p. 141.
```

Although in footnote citations to the publications of specialized agencies the corporate author is omitted, the name of the agency responsible for the publication should be brought out clearly in one of the items of the citation. The documents published by these agencies should be cited in a manner which will not allow them to be confused with documents issued by the United Nations (see pp. 75–77). The footnote reference should include the title of the document (italicized), the document or series number, the imprint, and the page or pages. Standard abbreviations may be used in place of the full name of the agency, except when the agency is known by its name rather than its initials, e.g., International Monetary Fund.

[1]<u>Nutritional Deficiencies in Livestock,</u> by Richard R. Allman and T.S. Hamilton, FAO Agricultural Studies, no. 5 (Washington: 1948), pp. 41-3.

> Food and Agricultural Organization may be given in full in the series note. This document may also be cited by personal author; see p. 31.

[2]<u>Indigenous Peoples, Living and Working Conditions of Aboriginal Populations in Independent Countries,</u> ILO Studies Reports, n.s., no. 35 (Geneva: 1953), p. 541.
[3]<u>Balance of Payments Manual,</u> Final Draft, International Monetary Fund (Washington: 1960), p.212.
[4]<u>Convention and Privileges and Immunities of Specialized Agencies. . .</u> U.N. Doc.: ST/LEG/4 (1953), Sales No. : 1953. X.I, p. 51.
[5]<u>Lists of Inter-Continental Telephone Communication Channels (Direct Circuits),</u> ITU Publication, no. 73 (2d ed.; Geneva: 1962), p. 17.

References to articles in periodicals issued by specialized agencies are cited in the form recommended for periodical articles on pages 86–87.

[1]F. Samuilenko, "Stabilization and Training of Manpower in the Forestry Industry of Byelorussia," <u>International Labour Review,</u> 83 (June, 1961), 533.
[2]"United States Book Exchanges," <u>UNESCO Bulletin for Libraries,</u> II (September, 1948), p. 328.

10. CONGRESSIONAL RECORD AND PARLIAMENTARY DEBATES

A footnote reference to debates and speeches in Congress should be a specific citation giving the number of the Congress and session, volume number, date, and page numbers. The reference is made by title (in italics) and, if possible, to the bound edition (see pp. 37–38).

¹<u>Congressional Record,</u> 79th Cong., 1st sess., 91:5 (June 14, 1945), 6085.

A fuller form of citation is sometimes found. The full form below should be used only when the subject, i.e., title, has not been brought out clearly in the text.

¹"Cooperatives - Bulwark of Democracy," Remarks of the Hon. Wright Patman, <u>Congressional Record,</u> 79th Cong., 2d sess., vol. 92, pt. 12 (July 5, 1946), pp. A4248-52.

Footnote references to speeches and debates in Parliament should give the precise title (see p. 47), series if necessary, volume, date, and column numbers. *Parliamentary Debates* are printed with two numbered columns to each page; citation is made to the column numbers. The abbreviations shown below are preferred to such shortened forms as *H.L.D., H.C.D.,* or *H.L. Deb.,* and *H.C. Deb.,* unless they are used in specialized studies.

¹<u>Hansard, Parl. Hist.,</u> XVI (1765-71), 497-501.
²<u>Parl. Deb.,</u> 4th series, 58 (1898), 560-61.
³<u>House of Commons Deb.,</u> 5th series, 419 (1945-46), 1759-65.
⁴<u>House of Lords Deb.,</u> 5th series, 138 (1945-46), 783-4.
⁵"Anglo-American Financial Arrangements," <u>House of Lords Deb.,</u> 5th series, 138 (Dec. 17, 1945), 783-4.

11. LAWS, STATUTES, ETC.

U. S. federal laws, etc.—Laws, court decisions, etc., are cited specifically in the footnotes (cf. p. 38). The extensive documentation required in legal publications demands compact forms of citation. Highly standardized practices of citation in the publication of legal literature have made it possible to use skeletonized abbreviations which would be impractical in other fields, where publishing practices lack uniformity. This shortened citation characteristic of the law is used extensively, however, in some other fields, when references to laws or court cases are necessary. The forms are not fixed, but the selection of items included in legal citation is uniform; the sequence of these items and the use of italics vary in different publications. The arrangement of items and use of italics in the examples below represent fairly common usage. The volume number (or code title) precedes the italicized title of the compilation, which is followed by the date, enclosed in parentheses; the subdivision or page reference is the last item; the imprint is customarily omitted.

¹58 U.S. Stat. at L. (1944), 284-301.

> For pages 284-301 of volume 58 of the *United States Statutes at Large*, published in 1944. Legal citation: 58 Stat. (1944), 284-301.

²U.S. Rev. Stat. (1878), sec. 388.

> For section 388 of the 1878 edition of the *U. S. Revised Statutes*. Legal citation: Rev. Stat. (1878) § 388.

³20 U.S. Code (1940), sec. 221.

> For section 221 of title 20 of the 1940 edition of the *Code of Laws of the United States*. Legal citation: 20 U.S.C. § 221.

⁴20 U.S. Code Supp. IV (1945), sec. 221.

> For section 221 of title 20 of the *Code of Laws of the United States*, cited in its fourth Supplement published in 1945. Legal citation: 20 U.S.C. Supp. IV (1945) § 221.

⁵Public Law No. 585, 79th Cong., 2d sess. (August 1, 1946). "Atomic Energy Act of 1946."

A Public Act should not be cited after it has been incorporated in the *Statutes at Large* (cf. p 39). The reference given just above should be changed to:

⁵60 U.S. Stat. at L. (1947), 755-75.

The title remains unchanged, and may or may not be included in the reference.

If the Constitution of the United States is mentioned in the text, give only the Article and section numbers in the footnote; otherwise cite in the footnote as follows:

¹U.S. Const. Art III, sec. 2.

> A clause number (e.g., cl. 2) may be added, if necessary.

²U.S. Const. Amend. XX, sec. 5.

State laws, etc.—Footnote citations to state laws (acts, session laws, statutes) should give the name of the compilation of laws italicized (usually abbreviated), the date in parentheses, the subdivision or pages (see also p. 41).

¹R.I. Gen. Laws (1938), chap. 238.
²Kan. Laws (1945), chap. 144.
³S.C. Acts (1946), p. 1304.
⁴Nev. Stats. (1945), p 458.

Codes or compilations of state laws are cited by name of compilation in italics; editor (if any), and date, both within parentheses; and section number.

¹*Iowa Code* (1946), sec. 109.
²*Ohio Gen. Code* (Page, 1937), sec. 3597.
³*Nev. Comp. Laws* (1929), sec. 5672.
⁴*Ind. Stats. Ann.* (Burns, 1933), sec. 10.
⁵*Cal. Civ. Code,* sec. 143.

California Codes are cited without the date or name of the editor.

When citing state constitutions, give the date of the constitution, if the original constitution has been superseded. If the specific provision cited is an amendment to the constitution, the date of its adoption should be given.

¹*Ariz Const.,* Art. 8, sec. 1.
²*New York Const.* (1939), Art. V, sec. 7.
³*Cal. Const.,* Art. 1, sec. 14 (Amend. 1928).

English statutes.—The items to be included in footnote citation to English statutes are: the name of the compilation, regnal year, name of the sovereign, calendar year, and subdivisions of the statute with appropriate number citation. Practice varies in the use of punctuation and sequence of the items. Many authors include the name of the Act as the first item in the citation; others cite it as the last item. It is found both within quotation marks and written in italics. Since the enactment of the Short Title Act (1896), every Act contains a short title by which it may be cited. Legal citations are likely to be more fixed and abbreviated than the examples below, which illustrate various styles of citation in forms full enough to be easily understood (cf. p. 48).

¹*Great Britain Statutes,* 1 Edward VIII, chap. 12 (1936).
²*Great Britain Statutes,* 1935, 25 & 26 George V, chap. 42. "Government of India Act, 1935."
³*Law Report Statutes* (1919), 9 & 10 George V, chap. 39.
⁴*Statutes of the Realm,* 25 & 26 Edward I. 1. "Confirmatio Cartarum."

Citation of early English statutes is often made without the calendar date, or the name of the compilation.

¹31 Henry VIII, 8.

12. COURT DECISIONS

The simple form of reference used in legal literature has been adopted in other fields for the citation of court decisions. The reference should give the name of the case in italics (i.e., names of plaintiff and defendant, or titles with prefixes such as *ex parte*, etc.), the date in parentheses, the volume number, name of the report, and page on which the case begins. Legal publications, however, do not italicize the name of the case. Practices in the arrangement of the date vary; sometimes it is given as the last item, but often it is omitted.

U. S. Supreme Court reports.—When citing decisions of the Supreme Court, citation of the official *United States Supreme Court Reports* (abbreviated simply U. S.) is always to be preferred. Often a duplicate citation of the case gives another edition of the Supreme Court decisions, the *Supreme Court Reporter* (abbreviated Sup. Ct.). A further duplication, citation of the case in the annotated *Lawyers' Edition* (L.ed.), is customary in legal publications. In duplicate citations the date need be given only once.

[1]Associated Press v. United States (1944), 326 U.S. 1.

If only the *Supreme Court Reporter* were available for citation, the reference would read:

[1]Associated Press v. United States (1944), 65 Sup. Ct. 1416.

When the reference includes citations of all three reports, it should read:

[1]Associated Press v. United States (1944), 326 U.S. 1; 65 Sup. Ct. 1416; 89 L.ed. 2013.

> United States is never abbreviated when it is part of the name of a case.

Citation of decisions of the lower federal courts, federal commissions, etc., should give the name of the court or commission in parentheses.

[1]In re Burrows (1946), 156 Fed. Rept., 2d ser., 640 (C.C.A.N.Y.).

> For vol. 156 of the *Federal Reporter,* second series, page 640. Case heard in the New York Circuit Court of Appeals.

[2]Lewis-Auburn Broadcasting Corporation (1946), 11 Fed. Reg. 12310 (F.C.C. Docket no. 7898, Oct. 19, 1946).

For vol. 11 of the *Federal Register,* page 12310, Federal Communication Commission Docket no. 7898.

State court reports.—Citation of official state reports is always to be preferred to other editions of reports, but duplicate citations of unofficial reports are customary (cf. p. 103).

[1]Williams v. Davis (1946), 27 Cal.2d, 746.

When state cases are cited by the name of the court reporter, the name of the state (jurisdiction) is included in the parentheses enclosing the date.

[2]State v. Caruso (Del. 1942), 3 Terry, 310.

The *Pacific Reporter* covers thirteen western states including California (cf. p. 42). The first illustration above may be cited:

[1]Williams v. Davis (1946), 167 Pac.2d, 189.
189.

If the *Pacific Reporter* is cited in addition to the official report, it is placed second in the reference.

[1]Williams v. Davis (1946), 27 Cal. 2d, 746; 167 Pac. 2d,

English court reports.—Detailed instructions in the citation of early English cases would be impractical here. The student who has occasion to cite sources other than the *Year Books* is referred to the *Guide to the Contents of the Public Record Office*,[3] and V. H. Galbraith's *Introduction to the Use of the Public Records*.[4] The *Year Books* may be cited variously as follows:

[1]Year Books, 8-9 Edward III (Rolls Series).
[1]Year Books, Trin. 8 Henry VIII, f. 3a.

Trin. for Trinity Term; f. for folio.

[1]Stonehouse v. Bodvil, Year Books 37 Henry VI, F. 9. pl. 3.

Pl. may be used for plea or *placitum*.

Early English citations are usually dated only by the regnal year. In order to calculate the calendar year, it is necessary to determine the coronation date and add the number of the regnal year. Examples of early citations, tables of regnal years, and the corresponding calendar years, etc., may be found in guides to legal literature.

---

[3] (London: H.M.S.O., 1963), 2 vols.
[4] (Rev. ed.; London: Oxford University Press, 1953), 112 pp.

Although the regnal year and name of the sovereign are still retained as part of English statute citation, these items are no longer given in case citation. The first case below is cited by reporter and also as included in *English Reports Reprints* (see also pp. 48–49).

¹Smeed v. Foord (K.B. 1859), Ellis & Ellis, 602.
¹Smeed v. Foord (K.B. 1859), 120 English Reprints, 1035.

Citation of the official *Law Reports* is always preferable to the unofficial reports. Since 1891, citations have been made uniformly, and the following form is recommended for general use.

¹Bailey v. Hookway, Law Reports, 1945. 1 King's Bench, 266.

> For volume 1, page 266, in the 1945 reports of the King's Bench Division. In the *Law Reports,* volume numbers are not cumulative; rather each year begins a new series with volume 1 for each court.[5] Legal citation:
> Bailey v. Hookway [1945] L.R. 1 K.B. 266.

Citation of unofficial reports follows the same form as American case citation, except that the initials of the court should be included in the parentheses enclosing the date.

¹Rex v. Darry (K.B., 1945), 61 Times Law Reports, 407.

Citations of unofficial collections of reports dealing with special branches of the law are made as follows:

¹The Bremen (1931), 18 n.s. Aspinwall's Reports of Maritime cases. 252.

> For volume 18, new series, of Aspinwall's Reports...page 252, citing *The Bremen,* 1931. Legal citation: The Bremen (1931) 18 (n.s.) Asp.Cas. 252.

SHORTENED FOOTNOTE CITATIONS

(For Later Reference, Etc.)

After the first full footnote reference to a publication, later citations may be in shortened form. The author's given names or initials may be omitted, unless citations have been made to more than one author of the same surname. The titles may be given in shortened form, but a sufficient number of key words should be in-

---

⁵ Abbreviations and other examples may be found in *A Uniform System of Citation* (10th ed.; Cambridge: Harvard Law Review Association, 1958), pp. 60–66.

cluded to facilitate identification. Such words as Introduction, Report, Handbook, etc., used singly are not sufficient to identify a title. Subtitles and facts of publication are omitted in the shortened forms. The title and other items of the reference may be eliminated by the proper use of *ibid., op. cit.,* and *loc. cit.* as explained below. The inclusion of the title of a book or article in a shortened footnote citation is essential if references have been made to more than one book by the same author. The names of periodicals may be abbreviated intelligibly (see p. 87). In long works, the full form of a citation should be repeated occasionally to avoid difficulties for the reader.[6] In place of using repetitions haphazardly as they become necessary, it is advisable to repeat the full form of the footnote with the first reference to a work in each subsequent chapter. The footnote references given below have been chosen solely to illustrate the bibliographical points discussed; the citations are not intended to convey a suitable selection of works on the basis of content or logical continuity.

Use of ibid.; op. cit.; loc. cit.—The abbreviation *ibid.* (i.e., *ibidem*, meaning "in the same place"), should be used for repeating the reference immediately preceding, provided the successive footnotes are on the same page or, in printed works, on any two pages which would lie open to the reader at the same time. *Ibid.* may be substituted for as much of the preceding reference as is needed—author's name, title, imprint, volume number, chapter, etc.—and should be followed by the new items (e.g., volume, section, paragraph, or page citation) which are necessary to complete the reference. A comparison of the shortened footnote illustrations below with the bibliographical and first full footnote illustrations will be helpful.

---

[6] Repetition of the footnotes may be avoided by numbering all citations and listing them consecutively at the end of the book, chapter, or article. The number of an individual citation may then be used to refer the reader to the "terminal bibliography." The use of footnotes and a formal bibliography, however, is a more satisfactory method of handling references, except in the scientific and technical fields where textual references and a "terminal bibliography" are used customarily. See Section III, 'Scientific and Technical References."

¹Norman Angell and others, Economic Principles and Problems (4th ed.; New York: Farrar and Rinehart, 1941), II, 123-4.
²Ibid., I, 218.
³H.L. Mencken, The American Language... (4th ed. rev. and enl.; New York: Knopf, 1938), p. 189.
⁴Ibid., Supplement II (1948), p. 17.
⁵David McCord Wright, "The Future of Keynesian Economics," American Economic Review, XXXV (June 1945), 293.
⁶Ibid., p. 297.

Op. cit. (i.e., *opere citato,* meaning "in the work cited") is used with the author's surname when repeated references to a book, article, or other publication occur (*a*) with interruptions by other footnotes on the same page, or (*b*) in footnotes falling on different pages. Obviously, the precise meaning ("in the work cited") prohibits the use of *op. cit.* if more than one book or article by the same author has been cited. The author's given name (or initials) is omitted, unless two authors of the same surname are cited. *Op. cit.* refers to a particular work of an author and should not be used to replace the name of a periodical.

Illustrations (continuing from above)

⁷Dexter Masters and Katharine Way, eds., One World or None (New York: McGraw-Hill, 1946), p. 33.
⁸Angell, op. cit., I, 34.
⁹Medical Care in the United States: Demand and Supply, 1939 (Chicago: American Medical Association [c. 1940]), pp. 123-7.
¹⁰Wright, op. cit., p. 305.
¹¹Atomic Energy Act of 1946, Hearings, 79th Cong., 2d sess., Senate, Special Committee on Atomic Energy (Washington: 1946), pp. 489-93.
¹²Ibid., pp. 221-2.
¹³Masters and Way, eds., op. cit., p. 34.
¹⁴Ibid., pp. 76-7.

Loc. cit. (i.e., *loco citato,* meaning "in the place cited") should be used in the place of *ibid.* or *op. cit.* when the repeated reference is to the exact page or passage previously cited. With *ibid.*, or with *op. cit.*, one may refer again to the work, but not to the specific pages given in a previous reference; when *loc. cit.* is used, the reference is not only to the work previously cited, but *also* to the same place or passage in that work. If *loc. cit.* is used immediately following the footnote to which it refers, the author's name, or other identification of the reference, need not be repeated (see illustrations 19 and 22 below, cf. 16 and 27). It should be noted that *loc. cit.* does not refer to the name of a periodical, but to an author's book, article, or other work. Neither *op. cit.* nor *loc. cit.* should be used for referring to a footnote in a previous chapter.

Illustrations (continuing from above)

[15] Medical Care in the U.S., p. 9.
[16] Fiorello La Guardia, "Why New York Should be the World Capital," P M (New York), Feb. 3, 1946, p. 3.
[17] Angell, op. cit., I, 143-4.
[18] Ibid., II, 57.
[19] Loc. cit.
[20] Annual Report of the National Science Foundation . . . 1962, 88th Cong., 1st sess., H. Doc. 4714 (Washington: G.P.O., 1963), p. 355.
[21] United States Government Printing Office Style Manual (rev. ed.; Washington: G.P.O., 1967), pp. 297-302.
[22] Loc. cit.
[23] C. B. Horne, "A Geographer Looks at Russia," Current History, n.s., 8 (Feb., 1945), 147.

The corporate author is customarily omitted in footnote references when the authorship of the work is shown by the title (21), series note (11, 20), or the imprint (9).

The use of *op. cit.* and *loc. cit.* should not be overdone. The first full footnote reference should be available to the reader without undue searching backward through shortened citations. If shortened footnote references are far removed from the first one that gives the full bibliographical information, the subsequent references should be written with at least a shortened title. The shortened title, however, should be inclusive enough to be unambiguous.

Illustrations (continuing from above)

[24] Atomic Energy Act . . . Hearings, pp. 235-6.
[25] Felix Frankfurter, "Benjamin Nathan Cardozo." Dictionary of American Biography, Supplement Two, XXI (1959), 194.
[26] Mencken, American Language, p. 318.
[27] La Guardia, loc. cit.
[28] Masters and Way, eds., One World or None, pp. 76-7.

²⁹Horne, op. cit., p. 144.
³⁰Ledoux, op. cit., p. 632.
³¹Wright, "Future of Keynesian Economics," pp. 299-301.

To shorten later footnote references, some authors adopt an abbreviation for standard works, e.g., *CHEL* for *Cambridge History of English Literature.* In the first full footnote reference the reader is informed of the use of the abbreviation by a specific notice such as: "hereinafter cited as *CHEL.*"

Abbreviations, symbols, etc.—In footnotes, and to some degree in the text of a manuscript (preferably in parenthetical matter only), various symbols, abbreviations, and Latin words are commonly used. Some writers, however, employ only the English equivalents and avoid the use of the traditional Latin words and abbreviations; but it is not necessary, and often is not practicable, to avoid some mixture of the two. A writer should, however, be consistent in his use of like terms for like situations, and not, for example, use "see above" in one place and *"vide supra"* in another. Specific page citations are preferred to abbreviations which indicate scattered references throughout several pages (for example: *et seqq.* or ff.). The following selected symbols, Latin words, and abbreviations are frequently employed and should be known to the student. Others may be found in any good dictionary.

SYMBOLS AND ABBREVIATIONS

	Used to denote an omission from a quotation, title, etc. This form of ellipsis may indicate omission of matter of any length, including the punctuation.
[ ]	Information enclosed in square brackets is supplied by the writer or editor; for example, p. [12] indicates that the page number is not printed on the page cited. Brackets may be used also for an editor's corrections and interpolations in direct quotations.
ante	before; cf. *supra*
art.	article
bk.	book; with a capital B, if preceding a roman (capital) numeral, e.g., Bk. II
bull.	bulletin
c.	copyright. In law citations, it is used for chapter.
ca.	(*circa*) about. Used of dates, e.g., *ca.* 1400.
cf.	(*confer*) compare. Sometimes cp. is used. One should take care to distinguish between "cf." and "see."

col.	column
comp.	compiler, compiled
diss.	dissertation
ed.	edition, editor, edited by
e.g.	(*exempli gratia*) for example
enl.	enlarged
et al.	(*et alii*) and others
et seq., et seqq.	(*et sequens* [singular], *et sequentes* or *et sequentia* [plural]) and the following. Used to indicate that the reference extends beyond the line or lines, or page or pages, cited. In general, f. or ff. are to be preferred.
f., ff.	and the following page, or pages; sometimes used for folio.
facsim.	facsimile
fig.	figure
fol.	folio
i.e.	(*id est*) that is
ibid.	(*ibidem*) in the same place
illus.	illustrator, illustrated
infra	below. "See below," the English equivalent of *vide infra*, is used as a cross reference to subsequent information in the work. Cf. *post*.
l., ll.	line, or lines, of verse. In a bibliography this abbreviation may be used for leaf, leaves.
loc. cit.	(*loco citato*) in the place cited; in the passage last referred to
MS, MSS	manuscript, manuscripts. Written without a period, and in capital letters.
n., nn.	note, notes
N.B.	(*nota bene*) note well; take notice
n.d.	no date
no.	number
n.p.	no place. Used in imprint.
n.s.	new series
numb.	numbered
op. cit.	(*opere citato*) in the work cited
p., pp.	page, pages
par.	paragraph
passim	Used in place of page references when the reference is to various passages; here and there; all through
pl.	plate
post	after. Used as a cross reference. Cf. *infra*.
proc.	proceedings
pt.	part; with a capital P if preceding a roman (capital) numeral, e.g., Pt. III
q.v.	(*quod vide*) which see; usually used to refer to parts within the same work
rev.	revised, revision
rep. rept. rpt.	} report

sec.	section
ser.	series
sic	so, thus. Inserted in square brackets after quoted material, to indicate that the quotation is literal, e.g., a misspelled word: frend [*sic*]; or an erroneous date, 1890–1810 [*sic*]; used without a period.
supra	above. "See above," the English equivalent of *vide supra*, is used as a cross reference to preceding information in a work. Cf. *ante*.
s.v.	(*sub verbo*, or *sub voce*) under the word; used in references to dictionaries and glossaries.
tr.	translator, translated.
trans.	transactions; sometimes used for translated
v., vs.	(*versus*) against
v., vv.	verse, verses
viz.	(*videlicet*) namely, to wit
vol.	volume; with a capital V if preceding a roman (capital) numeral, e.g., Vol. IV. May or may not be capitalized with arabic numerals.
vs., vss.	verse, verses. Cf. v., vv., above.

Capitalization and italics.—The abbreviations in the foregoing list will be found variously italicized and capitalized; there is no set practice among editors and publishers. With minor exceptions, foreign words and phrases should be italicized when used in a work in English. Throughout this manual, abbreviations of parts of a book or set of books (e.g., art., bk., pt., chap., vol., sec.) are written with a capital letter when they are followed by roman capital numerals, but with a small letter when followed by arabic numerals (e.g., Vol. II; vol. 2). Detailed rules of usage will be found in the style manuals listed on pp. 149–150.

Classical and literary references.—Footnote references to classical literature are conventionally made in shortened form, omitting the imprint and page numbers. The items cited include the author's name, the title of the work, and the subdivision of the work (e.g., book, chapter, section, canto, lines, etc.), as required by the arrangement of the work being cited. Some authors eliminate all commas except those indicating a succession of lines, parts, etc.; but in the illustrations given below, the author's name is separated from the title by a comma, and periods are used to separate the remaining items of the reference.

[1]Euclid, **Elements** I. 5.
    For Book I, Proposition 5.

²Aristotle, <u>Nicomachean Ethics</u> X. 13. 1, 5.
   For Book X, chapter 13, sections 1 and 5.
³Virgil, <u>Aeneid</u> VI. 126.
   For Book VI, line 126.
⁴Homer, <u>Odyssey</u> I. 267-83.
   For Book I, lines 267-83.

Although these examples illustrate the conventional form, it is often advisable to identify the numbered parts by the use of appropriate abbreviations (Aristotle, *Nicomachean Ethics,* Bk. X, chap. xii, secs. 1, 5), unless the work is intended for readers who are familiar with classical or literary citations (see footnote 1, p. 4).

Citations of works of literature which have been published in many editions are also given in a shortened form, omitting edition, imprint, and page references. This general reference enables the reader to locate the quotation in any edition. (Cf. footnote 1, p. 4, where the citation refers to a specific edition and gives page number.)

¹Pope, <u>Rape of the Lock,</u> Canto IV, line 34.
   Or: IV.34.
²Wordsworth, <u>The Prelude,</u> Bk. IX, lines 161-5.
   The abbreviations l. and ll. for line and lines are not used because of the possibility of confusion with the numbers which follow.
³Poe, "Ulalume," line 9.
⁴<u>Othello,</u> III, iv, 72-6.
   For Act III, scene iv, lines 72-6.

When an author's works are as well known as the plays of Shakespeare, it is not necessary to give the author's name in the footnote. It is customary to cite line references to the Globe Edition of Shakespeare's plays; other editions should be indicated (see p. 12).

Footnote references to Biblical passages are made simply by giving the name of the book, which is abbreviated but not italicized, and the chapter and verse numbers, separated by a colon. Unless otherwise indicated, Biblical references cite the King James Version.

¹Eccles. 6: 9.
²II Cor. 13: 15.

SCIENTIFIC AND TECHNICAL REFERENCES

As IN the social sciences, widely varying forms of reference will be found in other scientific and technical literature. These variations are not fundamental, but concern rather the sequence of different parts of the reference, capitalization, use of italics, etc. The general principles set forth in the preceding sections apply here; and the student should understand them, if only to recognize the reasons underlying variations in his own field. Citations in science sometimes require a different emphasis. Because of the rapid advances in certain fields of scientific inquiry, the date of publication often assumes an added importance, which is reflected by its prominent position immediately following the author's name. Footnotes are commonly omitted; instead, citations are grouped in a so-called "terminal" bibliography at the end of an article or chapter. The items of the terminal bibliography are usually given in abbreviated form. Even where shortened forms are employed, however, a knowledge of the full forms is essential if the modifications are to be made intelligently. In spite of the special problems which occur, many common necessities remain. Uniformity of style throughout one work, accuracy, and completeness of information are still the inescapable requirements, and the references given should meet them.

It is suggested that full references be made at the time one is collecting data for a paper. In the last revision of the manuscript, the forms may be altered to conform to the style required by the journal in which the contribution is to appear, or to departmental requirements if the written work is a dissertation.

References in scientific works are variously placed—listed in a terminal bibliography at the end of the work, in bibliographies at the ends of chapters, or given in footnotes throughout the work. The combination of footnotes and a formal bibliography occurs much less frequently, however, than in other fields. References to articles are customarily shortened, but references to books are likely to be made in a fuller form. In the biological sciences and medicine, references often note in detail the presence of plates, figures, and other illustrations.

The titles of books are found enclosed in quotation marks, or italicized, or written without any marks of emphasis. The names of periodicals are seldom italicized. Small capital letters are sometimes employed to accentuate either the author's name or the name of the periodical; but this refinement cannot be shown in typescript. It is recommended that in manuscripts, so far as is compatible with the bibliographical practice in the subject field, the titles of books and the names of periodicals be italicized, and that the titles of articles and chapters be placed within quotation marks. It will be observed in the illustrations below that, except for the first word, the titles of articles and books are often written without capital letters. The names of periodicals are abbreviated, but these abbreviations are given in widely varying forms. It is suggested that the student use only standard abbreviations, found in such works as *Chemical Abstracts* and the *Cumulative Index Medicus*.

Alternatives to footnote citation.—Various alternatives to the customary method of footnote citation are current in scientific and technical literature. The common aim of these alternatives is the elimination of the footnote from the bottom of the page. The terminal bibliography is used as compensation for this omission. Some device within the text directs the reader's attention to references grouped at the end of the article or chapter. These references are numbered consecutively, arranged alphabetically, or listed by the date of publication of the citation. The two methods described below take account of all except minor differences in the schemes used in connection with terminal bibliographies.[1]

Reference to a publication is made by mentioning the author's name in the text and following it immediately with a reference number in parentheses (square brackets in text which contains mathematical symbols), in the following manner:

```
No single experimental arrangement is possible, according
to Heisenbaum (1, p. 14), but the theory devised by Keith (2,
pp. 139-140) which predicted the outcome of the experiments
of Thorpe and Mann (5 vol. 2, p. 89) would appear to refute
his contention . . .
```

---

[1] Samuel F. Trelease, *How to Write Scientific and Technical Papers* (3d ed.; Baltimore: Williams & Wilkins, 1958), 194 pp.

The references are given corresponding numbers and arranged consecutively in a list at the end of the work. If the citations are to be specific, the customary total pagination or volumes must be replaced by exact page references. This compromise violates the principle which holds that the bibliography at the end of a work should list only complete articles or books. However, the practice is common in scientific publication and is entirely acceptable in that field. When specific references are grouped at the end of a work, the list should carry some such heading as "References" or "Literature Cited."

A variant of the first method maintains the principle of listing entire works in the bibliography. This scheme similarly follows an author's name in the text with a reference number, but it includes in the parentheses or brackets, with the reference number, the page citation. The reference number is underscored with a wavy line (indicating, in printing, **boldface** type) to differentiate it from the page numbers. The appended bibliography lists, with corresponding boldface numbers, entire works to which reference is made within the text. If this variant to the first method is used, the excerpt used in our first illustration will read:

```
No single experimental arrangement is possible, according
to Heisenbaum (1, p. 14), but the theory devised by Keith
(2, pp. 139-40) which predicted the outcome of the experi-
ments of Thorpe and Mann (5, vol. 2, p. 89) would appear to
refute his contention . . .
```

The second method emphasizes the date by mentioning the author's name in the text and following it immediately with the date of publication of the reference in parentheses. Citations to more than one publication by a given author in a single year are differentiated by writing the date, which serves as the index number to the references grouped at the end of the work, and adding lower-case letters, underlined, alphabetically in the order in which the references are mentioned (e.g., 1916*a*, 1916*b*). The date written in full is preferred to contractions such as ('16).

```
Progress was erratic. Malden's (1939) cell theory stimu-
lated the spectacular work of Canova (1941), but the later
applications by Geiser (1943) and Schultz (1944) disproved
any general application. Luckow's (1944) histological
classification and his subsequent (1948a) microscopic
```

studies of tissue gave a new impetus and led to a restatement of the theory by Luckow (1948b) himself.

These references are listed alphabetically by author at the end of the work. The works under each author are arranged chronologically by date of publication. Several works in any single year are designated by *a, b, c,* etc., as shown in the example. Occasionally, the references are listed chronologically, with the authors' names arranged alphabetically under each year.

The two methods described above embody the principal variations in the schemes of citation which employ the terminal bibliography. A discussion of the many divergences which exist with respect to sequence of items in the citation, inclusion or exclusion of title, use of lower-case roman numerals, or of boldface for volume number as distinguished from page number, would demand more space than could be justified in this general work.

The examples below have been taken from recognized journals in various fields of science and illustrate outstanding variations of the forms suggested in the main part of this manual. Following the name of the journal, examples of references to books are given first and periodical references are given second. All details in these examples, including capitalization, punctuation, and indentions, are reproduced precisely as found in the journals, except for differences in the size of type which convey varying emphases impossible to achieve in typescript. A wavy line underscoring a number indicates boldface type. Although the terminal bibliography is used commonly in scientific citation, it is not used exclusively. Footnote references have been included in the illustrations and are so identified. If footnotes are used, they are seldom accompanied by a bibliography.

Illustrations

American Mathematical Society—Transactions

   1. Richard Courant and Herbert Robbins, What is Mathematics? 3rd ed. Oxford University Press, New York, 1946.
   2. R.P. Dilworth, Note on complemented modular lattices, Bull. Amer. Math. Soc. vol. 46 (1940) pp. 74-76.

Association of American Geographers—Annals

[1]Hugh Seton-Watson, Eastern Europe Between the Wars, 1918-1941 (Cambridge, 1945), pp. 55-56.
[2]C. Warren Thornwaite: The Climates of North America according to a new classification, Geogr. Rev., Vol. 21 (1931), pp. 633-655.

These references appear in footnotes.

The Biological Bulletin

Wigglesworth, V.B., 1939. The principles of insect physiology. Dutton and Co., Inc., New York.
Kopac, M.S. and R. Chambers, 1937. The coalescence of living cells with oil drops. Jour. Cell. Comp. Physiol., 9:345-361.

Botanical Gazette

1. Bower, F.O. The ferns. Cambridge University Press, 1926.
2. Bausor, S.C. Effects of growth substances on reserve starch. BOT. GAZ. 104:115-121, 1942.

Institute of Radio Engineers—Proceedings

[1]Simon Ramo and John R. Whinnery, "Fields and Waves in Modern Radio," John Wiley and Sons, Inc., New York, N.Y., 1944, pp. 482-91.
[2]Marion C. Gray, "A modification of Hallen's solution of the antenna problem," Jour. Appl. Phys., vol. 15, pp. 61-65; January 1944.

These references appear in footnotes.

Journal of Biological Chemistry

1. Umbreit, W.W., Burris, R.H., and Stauffer, J.F., Manometric techniques and related methods for the study of tissue metabolism, Minneapolis, 37 (1945).
2. Rothen, A., J. Gen. Physiol., 24, 203 (1940).

Journal of Experimental Zoölogy

(1) Clark, W.M. 1928. The determination of hydrogen ions. Baltimore, Williams and Wilkins Co.
(2) Harris, M., 1943. The compatibility of rat and mouse cells in mixed tissue culture. Anat. Rec., vol. 87, pp. 107-117.

Journal of Geology

[1]A.S. Romer, Vertebrate Paleontology (2d ed.; Chicago: University of Chicago Press, 1945), pp. 590-91.
[2]D.M.S. Watson, "The Origin of Frogs," Trans. Roy. Soc. Edinburgh, Vol. LX, Part VII (1939-40), pp. 195-231.

These references appear in footnotes.

Journal of Physical and Colloid Chemistry

(1) Cohn, E.J., and Esdall, J.T.: <u>Proteins, Amino Acids and Peptides,</u> p. 520. Reinhold Publishing Corporation, New York, (1943).
(2) Fuller, C.S.: Chem. Rev. 26, 161 (1940).

Journal of the American Chemical Society

(1) Fieser, "Laboratory Manual of Organic Chemistry," 2nd ed., D.C. Heath and Co., Boston, Mass., 1941, p. 458.
(2) Cassie, <u>Trans. Faraday Soc.,</u> 41, 450 (1945).

These references appear in footnotes.

The Physical Review

[1] E.U. Condon and G.H. Shortley, <u>Theory of Atomic Spectra</u> (Cambridge University Press, Cambridge, 1935) pp. 73-78.
[2] B.E. Watt, Rev. Sci. Inst. 17, 334 (1946).

These references appear in footnotes.

Those who claim advantages for these alternative schemes to footnote citation maintain that there is less opportunity for error since the reference need be written only once. It appears, however, that the method which includes the specific page citation in the text reference offers a noticeable opportunity for error. Moreover, text references which include more than a superscript numeral make a spotty-looking page both in print and in typescript. It is suggested that these schemes of citation be used only for whatever practical advantages they offer in specialized publication, and when journal editors and publishers demand them. In works of general or semipopular appeal a fuller form of reference is to be preferred and should be made in footnotes and accompanied by a bibliography wherever possible (see Sections I and II above).

TYPING THE MANUSCRIPT

ANY manuscript submitted to a publisher, or to an examining committee for approval, should be complete in all respects and should present an attractive appearance. It gives the writer an opportunity to demonstrate his ability to assemble, interpret, and evaluate evidence perceptively and skillfully, and to present his conclusions in a logical and convincing manner. A manuscript of pleasing appearance bears evidence of painstaking attention, which inevitably contributes toward favorable reception of the work. On the other hand, an untidy manuscript carelessly assembled often handicaps favorable appraisal of the subject matter, since it is a common assumption that such a manuscript reflects untidy and careless habits of thought. In the last revision of a manuscript all references should be verified and checked for accuracy and, for consistency of form. The footnotes and bibliographical references constitute an integral part of a dissertation or other scholarly work. Faculty members concerned with the approval or rejection of a dissertation, readers in publishing houses, and editors of professional journals customarily investigate the references cited and thereby judge the accuracy of their interpretation, and their cogency and validity as supporting data. Experience has proved that faulty, inconsistent, or inaccurate references are an indication of mediocre if not inferior work in the manuscript as a whole. Formulae, mathematical equations, references to tables, and cross references to various parts of the text itself should be checked. Moreover, the manuscript should be read for consistency in all other details, such as spelling, capitalization, use of italics, etc. The ultimate responsibility rests with the author. "Inconsistency, in even minor mechanical detail, can throw distrust on an otherwise authoritative text."[1]

The preparation of a dissertation or manuscript in final form should be done by an experienced typist. The following instructions are designed to guide the writer if he is typing the manu-

---

[1] *Oxford Author's Style Book* (New York: Oxford University Press, 1943), p. 3.

script himself, or to assist him in instructing the typist intelligently concerning special problems. *Before* the dissertation is typewritten, the author should inform himself about special requirements by consulting the Dissertation Secretary or the Graduate Division Office as well as his graduate advisor. Specimen pages (pp. 135–147) illustrate acceptable practices. Forms suitable for submission to publishers are discussed at appropriate places in the following subsections.

Number of copies.—Usually at least two copies of the final draft of a dissertation (or thesis) are prepared. The original is customarily filed in the library of the institution granting the degree. A carbon copy is retained by the candidate for his personal use. Another copy may be required for deposit in his department.

Most universities make the original copy of a dissertation available for microfilming by University Microfilms, Inc. Consequently, the candidate will be asked to sign an agreement which permits his dissertation to be microfilmed and registered for research purposes. If the candidate wishes to copyright his dissertation, copyright arrangements should be made at the time this agreement is signed. It will be necessary to apply for copyright, pay the copyright fee, provide two positive film copies for deposit in the Copyright Office, and supply a copyright notice.

Some institutions accept a dissertation produced by one of the various photocopying processes. Such a copy must, however, be suitable itself for reproduction, be clearly legible, be reproduced on only one side of a durable paper that will not discolor, and the reproduced image should maintain its blackness.

Ordinarily, the Ph.D. dissertation must be accompanied by two copies of a 400 to 600 word abstract. Additional copies may be necessary for the use of the dissertation committee at the time of the examination. Care should be used in preparing the abstract because it will be published in *Dissertation Abstracts* (Ann Arbor.: University Microfilms, Inc., 1955–   ).

For those manuscripts intended for publication, the author should retain a carbon copy. It should contain all corrections made on the original, and carry exactly the same paging.

Materials: paper, etc.—The dissertation should be typewritten (if it is not printed) on plain, white, unperforated, and unglazed paper of standard typewriter size—eight and one-half by eleven inches. The original and first copies should be typed on a paper of substantial weight, at least sixteen-pound bond. A lighter-weight paper may be used for the second and third copies. The paper should be the same weight, color, and texture throughout, and under no circumstances should both sides of the paper be written on.

A dissertation should be typed in a uniform size of type throughout. Pica (10 letters to the inch) is usually preferred, but elite (12 to the inch) is also often acceptable. The keys of the typewriter should be thoroughly cleaned beforehand, and kept clean, in order to produce fine, distinct outlines. A black ribbon and black carbon paper are preferable to blue or other colors. Medium-finish carbon paper produces a blacker copy, but is more apt to smudge than hard-finish, which produces a somewhat grayish copy. The same finish should be used throughout the typing, and the carbon paper should be changed often enough to make sure that all copies are a clear and even black.

Erasures should be held to a minimum, and corrections should never be made by overstrikes. If erasures are necessary, they should be made cleanly and the erased part of the text should be replaced by typewriting of the same appearance as the rest of the page. A reasonable number of corrections (a few words) made in ink are acceptable to publishers, but the dissertation which is bound and preserved in a library should be free from ink corrections: otherwise a reader may be tempted to make further corrections in ink.

Margins.—A left-hand margin of one and one-half inches in dissertation manuscripts allows adequate space for binding. The other margins should be approximately of one width; one and one-quarter inches is an acceptable measure. But the top margins of the first page of text and the first page of each succeeding chapter or similar subdivision should be deeper than the top margins of other pages. Care should be taken to keep the bottom margins uniform in depth. Proper allowance should be made for footnotes, and they should not extend farther down the page than would the text. Some writers prefer to use a deeper margin at the bottom of the page than at the top, in imitation of a printed page. Most publishers require only one-inch margins, but wider margins are always a safe precaution and allow room for editing.

Typing instructions for dissertations.—The text should be typed with double spacing between the lines. The items of the bibliography and the footnotes may be typed with single spacing, but double spacing should be used between footnotes and between references in the bibliography (see p. 137; cf. p. 139). Three to four spaces are usually allowed between a chapter heading and the text below it. Centered headings and illustrations are set off with three or four spaces above and below, to make them stand out from the text.

The first line of each paragraph should be indented five to ten spaces, and single-spaced quoted material should be indented three to five spaces more. All indentations should be kept constant. Italics are indicated by underlining words or passages; wavy underlining indicates boldface.

In addition to the footnote acknowledgment, it is necessary to identify direct quotations, either by enclosing the quoted passage in double quotation marks or by arranging it with single spacing between the lines. Both styles are currently in use, but, in general, quotation marks are reserved for short quotations. The phase "short quotation" is ambiguous and variously interpreted. Some publishers and writers interpret it as a short sentence or short paragraph. Others specify the number of typewritten lines, which fluctuates from three to seven. It is suggested that the student observe the style common to his subject field, even if it requires that quoted material, regardless of length, be enclosed in quotation marks. When consecutive paragraphs of the same work are quoted, the quotes are used at the beginning of each new paragraph but only at the end of the last one. Double quotes are used for the main quotation, and single quotes for quotations within quotations.

Single spacing is also used to represent quoted material which would be set in reduced type on a printed page. In a typewritten manuscript, quotations of less than two full lines should not be written in this style, because the passage cannot be sufficiently differentiated from the text. It is recommended that single spacing be used only when the quoted passage is of at least five typewritten lines or two sentences, and that the quotation be separated from the text by three spaces. Single-spaced quotations should be indented three to five spaces at both text margins, unless the style of the subject field requires that these margins be kept flush with those of the text. If only one paragraph is quoted, the first line need not have the additional paragraph indention, but if the quoted extract exceeds one paragraph, the first line of each has the additional indention. Quotation marks are unnecessary.

Quotations of poetry which exceed one line, or which are entire short poems, are usually centered on the page, written with single spacing, and follow the line arrangement of the original. If poetry is placed within quotation marks and incorporated in the prose text, a diagonal line (virgule: /) should be used to indicate the line endings.

Typing manuscripts for publication.—Publishers require that all manuscripts be typed with *double spacing throughout* (including footnotes and quotations), in order to allow adequately for the necessary editorial comments, emendations, notations, and instructions to the printer.[2] The length of the pages should be kept uniform, both for appearance and to facilitate the printer's estimates. Twenty five or twenty-six double-spaced lines of pica type (12-point typewriter type), or twenty-seven or twenty-eight double-spaced lines of elite type (10-point typewriter type), make a well-proportioned page of about two hundred and fifty words.

Paging.—The pages of text in a manuscript are numbered consecutively in arabic numerals. The sequence may begin with the title page, or with the first page of text; publishers' practices vary. The preferred arrangement for dissertations numbers all preliminary pages, i. e., preface, table of contents, etc., in lower-case (small) roman numerals, and the regular sequence of arabic numerals begins with the first page of text proper. The numbers on the preliminary pages should be placed in the center of the bottom margin of the page, except for the title page, on which (though it

---

[2] *Harper & Row Author's Manual* (New York: Harper & Row, 1966), p. 6.

is included in the sequence) no number appears. The regular sequence of numbers should be placed in the upper right-hand corner, aligned with the right-hand margin of the text and separated from it by three spaces. The numbers are not punctuated.

Publishers prefer to have the manuscript paged in unbroken sequence from "1" on the very first page. This numbering reduces the possibility, at any stage in the publishing process, of inadvertently assembling the pages in an incorrect sequence. Each number in the sequence should be accounted for. If it becomes necessary to eliminate a page, the continuity of the sequence may be retained by adding the missing page number to the number on the preceding page (e.g., 15–16, or 33–36). Additional pages inserted later may be numbered 27A, 27B, etc., in conformity with the numbering on the preceding page. A note on the preceding page should account for the insertion; for example, on page 27, " 27A follows." Such adjustments are not acceptable in the final form of a dissertation, each page of which should carry but one number; later insertions should be avoided.

Title page.—Specimen copies of title pages, or instructions for typing them, are usually available from the Dissertation Secretary or the Graduate Division in institutions which require that dissertation title pages conform to a prescribed pattern. Before preparing the title page, the student should learn the local requirements, and then he should comply with them in all respects. In the absence of an institutional requirement, the exact title of the dissertation should be given in its fullest form, followed by the student's name written without degrees, professional titles, etc. A statement of the purpose of the dissertation, such as appears on the specimen title pages (pp. 135, 140), should include the name of the department, division, school, or college to which the dissertation is submitted and the name of the institution granting the degree. The date is usually the last item on the page. Committee signatures of approval may follow the title page on a separate sheet, but some institutions require that they appear on the title page itself.

All lines on the title page should be centered and properly spaced for balance and attractiveness. The title is typed in uppercase (capital) letters, and the rest of the page is typed in lower-case (small) letters, with capitals only where necessary. Only the necessary punctuation should be used; there should be no punctuation marks at the end of lines, if they can possibly be avoided; and underlining should be reserved to indicate italics. The title page should be kept as simple as possible. The title should indicate accurately and briefly the content of the study.

A manuscript submitted to a publisher should give the exact title and the author's name, with whatever degrees, etc., he wishes to appear after his name on the printed title page. These items are punctuated and placed precisely to indicate their relative importance. The publisher supplies the imprint and any other data that are to be included on the title page.

Preface.—Since dissertations do not carry the traditional dedication page, the preface follows the title page. The preface is the author's personal statement formally describing his use of materials. It may contain a discussion of the scope, limitations, or purposes of the study, or an evaluation of similar works, their omissions, and the consequent need of the present study, etc. The preface should be brief and should not include information which could be incorpated in the text or appendixes. Acknowledgments of any substantial assistance should be made here, rather than in a separate section. An "Acknowledgments" section is justified only when the acknowledgments occupy several pages. The author's appreciation should be expressed moderately and succinctly. Prefaces are written most often in the first person, but occasionally in the third. Usually they are signed, and the date and place of residence may then accompany the author's name or initials.

In manuscripts intended for publication, a preface, or a separate foreword, may be written by the editor (especially of a series) or by some distinguished scholar in the field.

Table of contents.—A fairly comprehensive table of contents is desirable in a dissertation, since an index is not customarily included. The chapter or other main subdivision numbers are written in upper-case roman numerals, the page numbers in arabic numerals. The columns of chapter numbers and page numbers are aligned vertically as shown on page 136 below. The chapter headings are written in upper-case letters. The relative importance of further subdivisions may be indicated by grouping them in some block or paragraph arrangement; in the absence of an index, a line-by-line enumeration is usually preferable. In printed works, several sizes of type can convey relative importance; but typed pages must show subdivisions by the use of capitals and small letters only. (Underlining is used exclusively to indicate italics.) A series of periods or dashes (printers call them "leaders") connects each heading with the page number to which it refers. A less detailed table of contents is adequate in a work which contains an index, but it should always be full enough to indicate the interrelationships of the divisions of the work (cf. specimens, pp. 136, 141). Punctuation other than that found in the headings themselves should not be used.

A half title may follow. It is not necessary in dissertations, however; and in manuscript copy for a printed work the publisher usually attends to its preparation.

Illustrations, etc.—Illustrative material, whether pictorial or other, should be listed on the page or pages immediately following the table of contents (see p. 142). Plates should be numbered consecutively, and the numbers should be differentiated from the page numbers by the use of upper-case roman numerals, or by some other device. If there are two categories of illustrative material, e.g., photographs and tables, each should be numbered individually and listed separately in the table of contents. The photographs might carry upper-case roman numerals, and the tables might be designated as "Table 1," "Table 2," etc. Illustrations which appear as part of the text may be referred to by page numbers or may carry their own numeration. Tables, statistics, and the like, should be typed in close approximation of the printed page; spacing and indention should indicate the relative importance of the items. All illustrative material should carry adequate descriptive legends. Footnote references to accompany charts or tables are made by using superior letters of the alphabet or symbols (see pp. 82–83, and specimen table, p. 144).

Photostatic copies of graphs, charts, tables, etc., may be incorporated in the carbon copies of a dissertation, but the originals should be included with the typewritten copy. Photostats, however, are generally unacceptable to publishers because they are rarely satisfactory for purposes of reproduction.[3]

Introduction.—The introduction may follow as a separate section, it may be incorporated as an integral part of the first chapter, or it may be treated as the first chapter and given a chapter number. If it is treated as part of the text, the regular sequence of arabic numbers begins here.

Headings.—Each heading of a chapter or subdivision thereof should be a brief expression of the main idea discussed in the part of the text to which it pertains, and should reflect a logical development of the subject. Chapter headings may be expressed in relatively general terms; lesser subdivisions, more precisely. Each chapter or main division should begin on a new page, with the number designation in upper-case roman numerals followed by the chapter title or heading in uppercase letters (e.g., XII. VALLEY FORGE). It is advisable not to begin a chapter with any sort of secondary heading, but rather to start with lines of normal text introducing the matter which *thereafter* is differentiated by appropriate centered headings or side headings. Headings should follow a simple and consistent plan employed uniformly throughout the work. Headings equivalent in importance should be typed in the same style and should occupy the same position in relation to the text, if the essential distinctions are to be evident. The main headings should be centered on the page and set off from the text by three spaces above and three below. Periods are not used at the end of lines. Underlining is used for side headings which need to be differentiated from the text.

Footnotes.—Footnotes are written in paragraph form, with the first line indented three to five spaces. They may be presented in the manuscript in three different arrangements. (1) The style recommended and used in this manual places the footnotes at the bottom of the page. (2) They may be grouped together at the end of a

---

[3] *Harper & Row Author's Manual*, pp. 72–87. Suggestions for the preparation of material for reproduction.

chapter, section, or work. (3) Another arrangement scatters the footnotes throughout the text, each note being typed, between two horizontal lines, immediately after the passage to which it refers.

In dissertations, the style of placing footnotes at the bottom of the page, in imitation of the printed page, is practiced far more widely than the other styles. This style is not difficult to type if a slight pencil mark is first made at the edge of the paper, to indicate the point at which the text should stop. The pencil mark is easily made, if about fifty sheets of paper are fanned slightly and a penciled line is drawn across the exposed edge of the sheets. This line is drawn two to three inches from the bottom of the paper, and the resulting pencil mark on the edge of each sheet is used as a guide. More or less space than is indicated by the guide may be needed, depending upon the length and number of the footnotes to be typed on a particular page. With a little experience, the proper amount of space can be gauged readily from the guide mark, without removal of the paper from the typewriter. After the last line of text has been typed and the carriage has been returned, the platen of the typewriter is turned once; a line about two inches long is typed in; the carriage is brought back to marginal position; the platen is turned twice; and the footnotes are typed with proper indention and single spacing (see specimen pages, pp. 137, 145).

The footnotes should be numbered consecutively from "1" throughout each chapter, section, or monograph of moderate length. See pages 82–83 for use of symbols, etc.

A long footnote which would continue to the succeeding page should be avoided. However, if a long footnote is necessary, the continuation should be indicated by extending across the full text area of the succeeding page the line which separates footnotes from text; the continued footnote is then completed, and the footnotes referring to the text on that page follow in regular style.

Footnotes are sometimes grouped at the end of a chapter or other section of the work. They are numbered consecutively throughout each such subdivision. A superscript numeral within the text informs the reader of the presence of the reference placed elsewhere in the work. The arrangement of notes in groups at the end of chapters and without text numerals is also used in textbooks, where the references are most often not specific citations but rather lists of related material for the guidance of the student. It is also used in semipopular works, both with and without text numerals, and here again it is usually a list of general references for the convenience of the reader. This style of writing footnotes is not recommended for dissertations (cf. pp. 113 ff.).

The style which scatters footnotes throughout the text simplifies the typing, but, in the opinion of most, also produces an unsightly page and unwarranted interruptions in the continuity of the text. It is seldom used in manuscript copy for printed works, but is occasionally found in dissertations. The style of placing footnotes at the bottom of the page, in imitation of the printed page, makes a more attractive page and is generally preferable. However, if the scattered style is adopted, the superscript numeral follows the word or passage to which it refers and is placed outside the punctuation. It may be typed by single-spacing (turning the platen of the typewriter once) after the line containing the superscript numeral; typing a line across the text area of the page; double-spacing; writing the footnote in regular paragraph form, preceded by the superscript numeral; single-spacing; typing another line the same length as the first one; double-spacing; and continuing the text. Sometimes a short line is used to set off the footnote, but the line across the page as described above is more effective and less choppy-looking. The footnotes should be numbered consecutively throughout the sections or chapters, etc. (see specimen page, p. 138).

Some publishers recommend that this style be used in manuscripts submitted for publication because it allows the printer to place footnotes on the proper pages with a minimum of inconvenience. In the galley proof, the footnotes remain scattered throughout the text, but when a galley of type is made up in pages, the printer can with small inconvenience transfer the footnotes to the bottom of the pages to which they refer. However, this advantage may also be gained with use of the first arrangement described, if the footnotes are omitted from the bottom of the pages in the publisher's copy of the manuscript. The footnotes are then numbered consecutively for each chapter or section and typed on separate sheets which follow the text sheets. *Publishers generally require that the footnotes be typed double-spaced.*

Specific instructions for the footnote forms are given in Sections II and III above; see also specimen pages which follow this section.

Bibliography.—The bibliography appended to a dissertation should include every reference which has been cited in the footnotes. The references are typed with hanging indention and single spacing; double spacing separates the individual references (see p. 139). This arrangement enables the reader to locate a reference at a glance and is recommended here; however, other arrangements are acceptable.

Publishers prefer to have the bibliography typed with double spacing throughout. An alphabetical arrangement of the items is adequate in short bibliographies, but bigliographies of more than two pages are often more useful if they are classified in some manner appropriate to the materials (see p. 146). The essential points of information to be noted in bibliographical references are described in Section I above; see also specimen page of bibliography (p. 139).

Appendixes.—An appendix contains supplementary reference materials that are considered necessary to the interpretation of the text. Much of the material found in appendixes is of questionable value, and could profitably be incorporated in the text, but long tables, detailed reports, and folded materials encumber the text and are difficult to handle if placed too near the text to which they refer. They are conveniently placed outside of the text in an appendix, where they do not intrude but are readily available to the reader. If the discussion and conclusions in the text are based upon data gathered by means of forms, questionnaires, or other inquiries which have been circulated by the author, copies of the blanks are usually included in an appendix. Other appropriate appendix materials are maps, charts, glossaries, constitutions, reports, etc. The question of what constitutes a proper appendix in any given work is dictated by the author's choice and organization of the materials to be included in the text.

Index.—An index is rarely necessary or advisable in a dissertation, but it is an important feature in a book of any consequence, and indispensable in a book of reference. A good index enables the user to locate information easily and should include a sufficient number of cross references to make the information readily available. Long sequences of page references lumped under an entry defeat efficient use of the index. Numerous page references should be broken up, supplied with pertinent headings and subheadings, and arranged in chronological, alphabetical, or any other sequence which may seem appropriate. The type of materials and the amount of detail will determine its form, indentions, etc. The index for a printed work is commonly prepared from page proofs. An index is usually printed in double columns, but a manuscript index intended for the publisher should be typed double-spaced, one column to the manuscript page. The page numbers in the index should be painstakingly compared with the numbers of the pages referred to; even one inaccurate page reference discourages the user and casts doubt upon the usefulness of the index as a whole. Instructions for the preparation of an index may be found in various leaflets and manuals issued by publishers.[4]

---

[4] Sina Spiker, *Indexing Your Book: A Practical Guide for Authors* (Madison: University of Wisconsin Press, 1964), 28 pp.

Specimen Pages
References and Index

NOTE: The format of this book does not accommodate both a well-proportioned type page and an ideal typewritten page as described in the instructions given on pages 119 ff. The specimen pages here following are neither as wide nor as deep as pages typed on standard typing paper (8½ x 11 in.) would be; nevertheless, they may serve to illustrate the main points discussed.

WORDSWORTH'S POLITICS

A Study in the Conservative Mind

Wallace W. Douglas

A thesis submitted to the faculty of
the Graduate School of Arts and Sciences
of Harvard University
in partial fulfillment of the requirements for
the degree of Doctor of Philosophy in the
Division of Modern Languages

1 9 4 6

[Specimen title page, no. 1]

TABLE OF CONTENTS

PART I: REFORMER

I.	Revolutionist as Whig	1
II.	"It Was My Fortune . . ."	39
III.	"A Poor District"	73

PART II: THE MIND OF A CONSERVATIVE

IV.	Business Man	156
V.	Rentier	179
VI.	"Put Money in Thy Purse"	203
VII.	"The Finest Brute Votes in Europe"	235

PART III: TORY IN A NEW WORLD

VIII.	Political Action	275
IX.	Defeat	314
	Bibliography	327

[Specimen table of contents, no. 1; cf. no. 2, p. 141]

of some sort in the world."⁷ Unlike Crabb Robinson, we do not have to regard his accepting the laureateship because

> The enemies will consider this as a proof of a grasping disposition. I know him to be a very generous man, tho' anxious on money matters, [sic] And yet I shall not be able to refute the insinuations which will be cast on him on account of this acceptance.⁸

We do not have to wish, as Barron Field did, that the poet would stop writing, because "his petition and Letters on Copyright showed too much anxiety to make a pecuniary advantage of the reaction in favor of his poems." We do not have to suspect the "little secularity of mind in our divine poet."⁹ We have forgotten that . . . .

---

⁷David A. Wilson, *Carlyle at His Zenith* (London: Kegan Paul, Trench, Trübner; New York: Dutton, 1927), p. 121.

⁸Edith J. Morley, ed., *The Correspondence of Henry Crabb Robinson with the Wordsworth Circle* (Oxford: Clarendon Press, 1927), I, 486.

⁹*Ibid.*, II, 591.

[Specimen page of text, no. 1; cf. no. 2, p. 138; no. 5, p. 147]

Years later she was still worrying, though for rather different reasons:

This leads my thoughts to the woful [sic] state of money & the 'Money Market' Every year we grow poorer—interest so low—Rents not paid &c. &c.![11]

---

[11] Edith J. Morley, ed., The Correspondence of Henry Crabb Robinson with the Wordsworth Circle (Oxford: Clarendon Press, 1927), I, 219.

All his mature life, because he had never been a money-maker, and because his income never exceeded his wants, Wordsworth was plagued by a fear of not being able to meet the expenses of his family - of their education, of their illnesses, of what he conceived to be their "narrow circumstances."[12]

---

[12] Ernest De Selincourt, ed., The Letters of William and Dorothy Wordsworth: The Later Years (Oxford: Clarendon Press, 1939), II, 794; III, 1165, 1369. Middle Years, II, 832, 862. Morley, op. cit., I, 179.

[Specimen page of text, no. 2; cf. no. 1, p. 137]

Armitt, Mary L. Rydal . . . Ed. by W.F. Rawnsley. Kendal, [Eng.]: T. Wilson, 1916. 727 pp.

Ashley, William J. The Economic Organization of England. Lectures delivered at [the Colonial Institute] Hamburg. London, New York: Longmans, Green, 1914. 213 pp.

Aspinall, Arthur, ed. The Letters of King George IV, 1812-1830. Introduction by C.K. Webster. Cambridge, [Eng.]: University Press, 1938. 3 vols.

---- Lord Brougham and the Whig Party. University of Manchester Publications, Historical Series, No. XLVII. Manchester: University Press, 1927. 332 pp.

Bailey, John, and George Culley. General View of the Agricuture of the County of Cumberland . . . London: MacRae, 1794. 63 pp.

Baily, Francis. An Account of the Several Life-assurance Companies Established in London . . . 2d ed. London: Richardson, 1811. 49 pp.

[Barrett-Browning, Robert, ed.].The Letters of Robert Browning and Elizabeth Barrett-Barrett. New York, London: Harper, 1899. 2 vols.

[Specimen page of bibliography, no. 1, arranged alphabetically; cf. no. 2, p. 146]

INTERRELATIONSHIPS OF

THE AGENCIES CONTROLLING

SAN FRANCISCO BAY

By

Peyton Hurt

DISSERTATION

Submitted in partial satisfaction of the

requirements for the degree of

DOCTOR OF PHILOSOPHY

in

the Department of Political Science

in the

GRADUATE DIVISION

of the

UNIVERSITY OF CALFORNIA

Approved:

....

....

....

                Committee in charge

Deposited in the University Library......................
                                      Date Librarian

[Specimen title page, no. 2; cf. no. 1, p. 135]

TABLE OF CONTENTS

Chapter	Pages
I. INTRODUCTION	1-16
Control over the movement of goods and and persons	4
Control over navigation	8
Control over navigable waters	10
Interrelationships between the agencies of control	14
II. INTERRELATIONSHIPS OF THE AGENCIES CONTROLLING IMPORTS	16-58
REQUIREMENTS OF CUSTOMS ENTRY	17
The Customs Service and consular officers	19
COLLECTION OF DUTIES AND TAXES ON IMPORTS	21
The Customs Service and the Bureau of Internal Revenue	22
ADMISSION OF PLANTS AND PLANT PRODUCTS	25
The state Plant Quarantine Service and the federal Plant Quarantine and Control Administration	26
The federal Bureau of Plant Industry and the state Seed Inspection Service	28
The Customs Service and the Plant Quarantine Service	29
The Customs Service and the Bureau of Plant Industry	33
ADMISSION OF FOOD AND DRUGS	35
Control by consular officers	36

iii

[Specimen table of contents, no. 2; cf. no. 1, p. 136]

LIST OF CHARTS

		Facing page
I.	Interrelationships of the agencies controlling imports	19
II.	Interrelationships of the agencies controlling exports	60
III.	Interrelationships of the agencies controlling the movement of persons	75
IV.	Interrelationships of the agencies controlling navigation	94
V.	Interrelationships of the agencies controlling navigable waters	137

[Specimen list of illustrative material]

CHAPTER IV

INTERRELATIONSHIPS OF THE AGENCIES CONTROLLING THE MOVEMENT OF PERSONS

The regulation and control over the movement of persons to and from the United States by way of San Francisco Bay is entirely a federal function. Most of the regulations are concerned with the admission of persons into the United States and are under the immigration laws,[1] which are designed primarily to regulate the entrance of persons of alien nationality, but which include within their jurisdiction all persons arriving from abroad. These laws specify what classes of aliens shall be excluded and provide for the examination of intending immigrants prior to their departure from foreign ports as well as upon their arrival at ports in the United States. They forbid persons to bring into the United States any aliens not lawfully permitted to enter.

---

[1] 39 Stat. L. (1917), 874; 43 Stat. L. (1924), 153, as amended by 45 Stat. L. (1928), 1009; and 45 Stat. L. (1929), 1512, 1551.

[Specimen page of text, no. 3]

SEIZURES FOR VIOLATIONS OF THE CUSTOMS LAWS DELIVERED TO THE COLLECTOR
OF CUSTOMS AT SAN FRANCISCO DURING THE FISCAL YEAR ENDED JUNE 30, 1930

| COMMODITIES SEIZED | AGENCIES BY WHICH SEIZURES WERE MADE |||||||| TOTAL ||
| | CUSTOMS || COAST GUARD || OTHER AGENCIES* || TOTAL ||
	Seizures	Value	Seizures	Value	Seizures	Value	Seizures	Value
Narcotics	11	$39,222.54			1	$1,500.00	12	$40,722.54
Liquors	275	4,685.15	3	$36,074.00			278	40,759.15
Merchandise	297	13,764.39					297	13,764.39
Boats			2	8,000.00			2	8,000.00
Automobiles					8	5,700.00	8	5,700.00
Grand total							597	$111,945.93

SOURCE: Data supplied by the customs officer in charge of seizures.

*The seizure of narcotics valued at $1,500 was delivered by the San Francisco police, and the 8 seizures of automobiles valued at $5,700 were delivered by agents of the United States Bureau of Narcotics.

[Specimen table]

Netherlands, Germany, Norway, Denmark, Poland, Italy, and Czechoslovakia.[6]

The annual report of the Public Health Service for 1930 states that this system of medical examination of applicants for immigration visas has proved so satisfactory that it is proposed to extend it to additional countries as soon as trained medical officers are available for this purpose.[7] The Commissioner General of Immigration has likewise indicated his satisfaction with the work of technical advisers stationed at consulates abroad, and he briefly summarizes the working of the system as follows:[8]

> Immigration inspectors possessing a comprehensive knowledge of the immigration laws, expert in the art of developing essential facts

---

[6] The Immigration Work of the Department of State and Its Consular Officers (Washington: 1929), p. 4; see also Annual Report of the Surgeon General of the Public Health Service, Fiscal Year, 1930 (Washington: 1930), p. 187.

[7] Ibid., p. 176.

[8] Annual Report of the Commissioner General of Immigration to the Secretary of Labor, Fiscal Year, 1930 (Washington: 1930), p. 17.

[Specimen page of text, no. 4]

BIBLIOGRAPHY

a. Primary Sources

Laws, statutes, court reports, etc.
  California
    Constitution.
    Penal Code.
    Political code.
    The Pacific Reporter.
    Statutes.

  United States
    Constitution.
    The Federal Reporter.
    Revised Statutes.
    Statutes at Large.
    Supreme Court Reports.

California. Board of State Harbor Commissioners for the Port of San Francisco. Tariff Charges, Dockage, Tolls, Demurrage, and Rentals for the Port of San Francisco, and Rules and Regulations for the Operation of the State Belt Railroad and State Grain Terminals. No. 2. Sacramento: California State Print. Off., 1925. 52 pp.

U.S. Bureau of Animal Industry. Regulations Governing the Meat Inspection of the United States Department of Agriculture. Order, no. 211. Rev. ed. Washington: Gov't. Print. Off., 1922. 107 pp.

[*b., Secondary Sources,* follows and is arranged alphabetically; see p. 139 for specimen bibliography arranged alphabetically.]

[Specimen page of bibliography, no. 2]

[*146*]

is interested here primarily in the form of the transform and can disregard consideration of integration in the complex plane, it is not necessary to state precisely how large p needs to be.

As a consequence of the uniform convergence of the integral the following is true:*

$$\lim_{p\to\infty} X(p) = \lim_{p\to\infty} \int_0^\infty e^{-pt} x(t)\,dt = \int_0^\infty \lim_{p\to\infty}\left[e^{-pt} x(t)\right] dt = 0.$$

X(p) is defined as the transform of a function of the class x(t). This fact is useful in testing a function of p to determine whether or not it is a transform; but, while necessary, it is not sufficient.†

It will now be shown that the following functions of x(t)

a) $t^n x(t)$ $n > 0$

b) $\int_0^t x(t)\,dt$

c) $e^{bt} x(t)$

b any complex constant
are all of the class x(t). All three satisfy the conditions

---

*G. Gibson, *Advanced Calculus* (London: Macmillan, 1931), p. 441.
†Doetsch, *op. cit.*, p. 49.

SELECTED LIST OF REFERENCES

[Throughout this manual a variety of footnote citations have been made for illustrative purposes. These citations are valid and pertinent to the text, but it has not been considered necessary to include all of the works cited in this selected list, which is intended as a general guide to other manuals of style, works illustrating bibliographical form, etc.]

American Institute of Chemical Engineers. *Guide for Writers and Speakers ... Style Manual...* New York: A.I.C.E., [1965]. 24 pp.

American Institute of Physics *Style Manual for Guidance in the Preparation of Papers* . . . 2d ed. rev. New York: A.I.P., 1963. 42 pp.

American Psychological Association. *Publication Manual* . . . [Washington, D.C.: A.P.A., c.1957]. 70 pp.

Appel, Livia. *Bibliographical Citation in the Social Sciences and Humanities; A Handbook of Style* . . . 3d ed. Madison: University of Wisconsin Press, 1949. 32 pp.

Collins, Frederick H. *Authors' and Printers' Dictionary* . . . 10th ed. rev. London: Oxford University Press, 1956. 442 pp.

Conference of Biological Editors. Committee on Form and Style. *Style Manual for Biological Journals*. 2d ed. Washington, D.C.: American Institute of Biological Sciences, [c.1964]. 117 pp.

Cross, Louise M. *The Preparation of Medical Literature*. With a Chapter, "Charts and Graphs," by Shirley Baty. Philadelphia: Lippincott, [1959]. 451 pp.

Esdaile, Arundel. *A Student's Manual of Bibliography*. 3d ed. revised by Roy Stokes. London: Allen & Unwin, 1954. 392 pp.

Fieser, Louis F. and Mary Fieser. *Style Guide for Chemists*. New York: Reinhold, 1960. 116 pp.

Foster, John. *Science Writer's Guide*. New York: Columbia University Press, 1963. 253 pp.

Gensler, Walter J. and Kinereth D. Gensler. *Writing Guide for Chemists*. New York: McGraw-Hill, 1961. 149 pp.

Harper & Row Author's Manual. New York: Harper & Row, 1966. 141 pp.

Hockett, Homer C. *Introduction to Research in American History*. 2d ed. with corrections and appendix. New York: Macmillan, 1950. 179 pp.

Institute of Electrical and Electronic Engineers. "Information for IEEE Authors," IEEE Spectrum, II (August, 1965), 111–15. Reprints available from I.E.E.E.

Kinney, Mary R. *Bibliographical Style Manuals: A Guide to Their Use in Documentation and Research*. Assoc. of College and Reference Libraries, Monograph no. 8. Chicago: A.C.R.L., 1953. 21 pp.

McGraw-Hill Book Company. *The McGraw-Hill Authors Book*. New York: McGraw-Hill, 1955. 88 pp.

Selected References

"Manual for Authors of Mathematical Papers," American Mathematical Society *Bulletin*, LXVIII (Sept. 1962), 429–44.

A Manual of Style. 11th ed. rev. Chicago: University Press. [1952]. 522 pp.

Modern Language Association. *The MLA Style Sheet*. Compiled by William R. Parker. Rev. ed. New York: M.L.A., 1965. 30 pp.

National Education Association. *NEA Style Manual for Writers and Editors*. Washington, D.C.: N.E.A., 1962. 76 pp.

New York Times. *Style Book for Writers and Editors*. Edited and revised by Lewis Jordan. New York: McGraw-Hill, [1962]. 124 pp.

Nicholson, Margaret. *A Manual of Copyright Practice for Writers, Publishers, and Agents*. 2d ed. New York: Oxford University Press, 1956. 273 pp.

Orne, Jerrold. *The Language of the Foreign Book Trade; Abbreviations, Terms, Phrases*. 2d ed. Chicago: American Library Association, 1962. 213 pp.

Postell, William D. *Applied Medical Bibliography for Students*. Springfield, Ill.: Thomas, [1955]. 142 pp.

Price, Miles O. *Practical Manual of Standard Legal Citations*. 2d ed. New York: Oceana Publications, 1958. 122 pp.

Spiker, Sina. *Indexing Your Book; A Practical Guide for Authors*. Madison: University of Wisconsin Press, 1964. 28 pp.

Seeber, Edward D. *A Style Manual for Authors*. Bloomington: Indiana University Press, [1965]. 96 pp.

Trelease, Sam F. *How to Write Scientific and Technical Papers*. 3d ed. Baltimore: Williams & Wilkins, 1958. 194 pp.

A Uniform System of Citation. 10th ed. Cambridge: The Harvard Law Review Association, 1965. 124 pp.

United Nations. Dag Hammarskjold Library. *Bibliographical Style Manual*. Bibliographical Series, no. 8. ST/LIB/SER.B./8. 1963. 62 pp. Sales No. 63.I.5.

U.S. Department of Agriculture. *Bibliographic Style: A Manual for Use in the Division of Bibliography of the Library*. Bibliog. Bull., no. 16. Washington, D.C.: G.P.O., 1951. 30 pp.

U.S. Geological Survey. *Suggestions to Authors of the Reports of the United States Geological Survey*. 5th ed. Washington: G.P.O., 1959. 255 pp.

U.S. Government Printing Office. *Style Manual*. Rev. ed. Washington: G.P.O., 1967. 512 pp.

Williams, Cecil B. and Allen H. Stevenson. *A Research Manual for College Studies and Papers*. 3d ed. New York: Harper, 1963. 212 pp.

Yale University Press. *A Manual for Authors*. New Haven: Yale University Press, 1964. 38 pp.

INDEX

[This index lists only those items which are discussed in the printed text. It does not include the illustrations (in typewriter type) of bibliographical references and footnote citations which are intended to interpret the parts of the text which they follow and to illustrate the recommended forms of citation. The reference works named in the text are pertinent as cited, but they are cited primarily to illustrate various forms of footnote citation, and likewise are not included in the index. The organizations (e.g., International Institute of Intellectual Cooperation) which have been mentioned in the text, often parenthetically, but not discussed, are omitted.]

Abbreviations: capitalization of, 111; consistency in use of Latin, or English equivalents, 109; imprint, 12–14
────── In citation of: Biblical, classical, and literary works, 111–112; International Court of Justice, 98; manuscripts, 20, 89–90; *Parliamentary Debates*, 100; Parliamentary publications, 43–44; periodical articles, 87, 96, 106, 114; scientific works, 114; specialized agencies, 75, 99; standard reference works, 109; U. N. documents, 68, 69, 71, 96, 97, 99 (*see also* U.N. descriptive symbols); U.S. Congressional publications, 33–34; unpublished letters, 89–90
────── List of, 109–111
────── Uses of: in imprints, 13 (books); 33–34, 43–44 (government documents); in legal citation, 100–101, 102, 103, 105 n.; scientific references, 113–114
Accuracy, need for, 3–4, 81, 82, 113, 119
Acknowledgments: in preface, 125; of sources used, 1, 82, 122
Acronyms (U.N.), 63
A.L.S., 90
Alternate form of document reference, 37
Alternatives to footnote citation. *See under* Science, citations in *Annexes, Official Records*. See *Official Records* (U.N.)

Anonymous works, citation of: books, 9 (author's name known); 9, 18 (author's name unknown); encyclopedia articles, 26; periodical articles, 24
Appendix, 80; proper materials for, 130
Arabic numerals, uses of: in scientific references, 114–115; in U.N. document symbols, 62; with abbreviations, 111
────── To indicate: edition, 12; footnotes, 82, 84, 128; page numbers in manuscripts, 123, 125, 128; volume numbers in references to books, 15, 85–86; encyclopedic works, 26; periodical articles, 21, 23, 24. See also specimen pages 136, 141, 142
Articles, bibliographic references to: in encyclopedic works, 26–27; in newspapers, 25–26; in parts or sets of books, 27–28; in periodicals, 20–24; recommended forms for, 22–23
────── Footnote citation of: encyclopedic works or parts of books, 88–89; in newspapers, 87–88; in periodicals, 86–87; in scientific literature, 113–116; shortened forms of, 105–109; recommended forms for, 87 ff.
Asterisk, 80. *See also* specimen page 147
Atomic Energy Commission (U.N.), document symbol for, 62; citation of its *Official Records*, 68

Index

Author position (entry) in references, 15, 18, 24, 31

Author's name: forms used, 7; pseudonyms, 9; titles of nobility, 8; with "and others," 7. *See also* Anonymous works; Corporate author; Editor; etc.

——— Form of, used on manuscript title pages, 124, 125 *and* 135, 140

———In bibliographical references: books, 7–9; encyclopedic works, 26–27 government documents, 29–31, 35; manuscripts, 19; newspapers, 25; parts of books or sets, 28; periodical articles, 20, 23, 74; repetition of, 16 n.; written in inverted order, 7, 16, 20, 23, 28; written in regular order, 7

——— In footnote citations: government documents, 30–31, 96 (U.N.); not repeated if given in text, 83, 86, 96; of classical and literary works, 111, 112; of manuscripts and letters, 89; shortened forms of (using *ibid., op. cit., loc. cit.*), 106–108; use of "in his" with, 88; written in regular order, 84, 85

——— In scientific and technical references, 113, 114, 115; examples of, in scientific journals, 116–118

Author's responsibility: for acknowledgment of sources used, 1, 81–82, 125; for uniformity of all details in manuscript, 3, 119

Bible, references to, 112

Bibliographical references, 1–6; differentiated from footnote citations, 1–2, 81; integral part of manuscript, 119; modified forms of, for footnote citation, 3, 84, 113

Bibliographical terms, foreign, 7

Bibliography, the: absence of, 81, 84, 90–91; annotated, 5; arrangement of, 5, 113, 130; contents of, 1, 5, 130; differentiated from footnote citation, 1–2, 81; hanging indention, 16, 23, 130; in dissertations, 130; in textbooks, 5, 128–129; location of, 5, 106, 113, 135; repetition of author's name, corporate author, title, 16 n., 29; typing instructions for, 15–16, 23, 122, 130. *See also* Terminal bibliography. (*And see* specimen pages 139, 146)

Biological sciences, references in, 113, *and* journal examples, 117

"Blue Books." *See* Parliamentary publications *under* Gt. Brit.

Boldface type, 115, 116, 122, *and* examples, 117–118

Books: bibliographical references to, 6–18 (recommended form, 15–16)

——— Footnote citations to, 84–86 (recommended form, 85); classical and literary citations, 111; shortened forms of, for later use, 105–109

——— Parts of, bibliographical references to, 27–28; footnote citations to, 88

——— Sets of, with collective title, 28

——— Scientific references to, 113–114

Books, foreign, 7

Brackets. *See* Square brackets

British government documents, laws, statutes, etc. *See* Gt. Brit., government documents, etc.

Capitalization: of abbreviations (foreign words and phrases), 111; of prepositions and conjunctions in titles, 10, 15; of scientific references, 114, 116; of subtitles, 10; of titles of books, 10, 15–16; government documents (footnote citation), 91; of periodical articles, 20, 21, 23; of vol., p., pp., 15, 21, 22–23, 84, 111. *See also* "Symbols and Abbreviations," 109–111

—— In dissertations and other manuscripts: headings, 127; need for consistency of, 119; of title pages, 124–125; of tables of contents, 125–126. See also specimen pages 135 ff.

Carbon copies: for dissertations and manuscripts for publication, 120; paper for, 121

Chapter: in legal citation, 41, 101; references grouped at end of, 5, 81, 113, 127–128; usually omitted in references, 86, *but see* 27–28

Chapter headings (and numbering) in manuscripts, 125–126, 127

Charter (U.N.) See under United Nations

Charters, city, 42

Charts: located in appendix, 130; notation of, in references, 15; use of symbols in, 82–83 *and* 144, 147

Citation, lack of uniformity in, 2, 3, 65 (U.N.), 78–79 (regional organizations), 113–114 (science), *but see* legal citation, 100

Citation, principles of, 1, 3, 6, 65, 113

Classics, footnote citation to, 111–112

Codes (law), citation of: city, 42; state, 41, 102; U.S., 38, 100–101

Collaborators, 7

Collections cited by compiler, editor, etc., 8, by reporter (law), 39, 48

Colon, 12, 85 (imprint), 10 (subtitle), 24 (separating volume and page number)

Column numbers, notation of, in references to: League of Nations publications, 50; newspaper articles, 25; *Parliamentary Debates*, 100

Command Papers, 44. See also Parliamentary publications *under* Gt. Brit., government documents

Compiler, 8

Congress, notation of number of, in references, 32, 33, 38, 39

Congressional publications: bibliographical references to, 32, 33–34, 36; citation by personal author, 30–31, 91; footnote citations of, 90–91; serial numbers, 32–33. See also Corporate author

Consistency, need for, in citation, 3, 8, 37, 111, 114, 119

Constitutions, citation of, 51–52 Gt. Brit.); 41–42 (state), 39, 101 (U.S.). See also Appendix; Charters, city

Copyright, 14, 120 (dissertations)

Corporate author, 17

—— In bibliographical references to: books, 17–18; British documents, 45; foreign documents, 49; International Labor Organization, 54; League of Nations, 52; state and municipal documents, 41; United Nations, 66; U.S. government documents, 29, 30, 36

—— Omission of, in footnote citations to: books, 86; League of Nations, 92; U.N. Preparatory Commission, 95; U.N. publications, 96; U.S. national, state, and municipal documents. See also 108. Cf. Author's name; Personal author, citation of, in document references; Title entry

Correction of typing errors, 121

Court reports: bibliographical references to, 39–40 (U.S.), 42 (state), 48–49 (English); footnote citations to, 103 (U.S.), 104 (state), 104–105 (English)

Court reporters, 39 (U.S.), 48 (English)

Covenant. See *under* League of Nations

Cross references, 9, 19; use of footnotes for, 81, 119

Dagger, 82 *and* specimen page, 147

Dashes, uses of: to indicate work in progress, 14; to indicate repetition of author's name, title, 16 n., 141; corporate author, 29; as

leaders, 126 *and* 141, 142
Date in references to: books, 12, 14; cases (law), 100–101; Congressional publications, 32, 99; *Congressional Record*, 38, 99; English law (regnal year), 48; interviews, 89; *Parliamentary Debates*, 47, 100; U.S. and state laws, statutes, etc., 38, 41, 100–101; unpublished dissertations, 18, 89, 135, 140; unpublished letters and manuscripts, 19, 89
Date of publication: copyright date substituted for, 12, 14; emphasized in scientific references, 113, 115–116; in periodical references (date of issue,) 21–22, 24, 87; in newspaper references, 24; of encyclopedias, 26; replaced by n.d., 14; of U.N. documents, 66, 67, 68. See *also* Imprint
Dedication in manuscript, 125
Degrees, academic, 120; omitted in references and on dissertation title pages, 7, 124; on title pages of manuscripts for publication, 125
Departmental publications (U.S.): bibliographical references to, 35–37; cited by corporate author, 29; cited by personal author, 36–37, 91; footnote citations of, 91; serial numbers of, 32–33
Descriptive legends, 126
Dissertation Abstracts, 120
Dissertations: Candidates to consult Graduate Division or Dissertation Secretary, 120; number of copies, microfilming, and copyright, 120. See *also* Typing the manuscript
Documentation, 1, 81, 119

Edition, citation of, in bibliographical references to: books, 12; *Congressional Record*, 37–38; encyclopedias, 26 *and* n.; government documents, 31, 35, 43; laws, statutes, etc., 38, 39, 41, 45; newspapers, 25
—— In footnote citation of: Bible, 112; books, 84, 85, 86; Classical and literary works, 112; *Congressional Record*, 99; laws, statutes, etc., 101 n., 103, 104
Edition, reprint, 14
Editor, citation of, in: books, 8, 12, 28; dictionaries and encyclopedias, 18, 26; English court reports, 48; special editions, 12; state codes, 41, 102
Editorials (newspaper), 25
Elite typewriter type, 121, 123
Ellipses, 31, 109; used in quotations, 75; used in titles, 33, 66
Encyclopedic works, articles in: bibliographical references to, 18, 26–27; footnote citation to, 88
English court reports, government documents, statutes, etc. See *under* Gt. Brit.
Equations. See Mathematics, citation in
Et al., 7, 110
Executive departmental documents (U.S.). See *under* U.S.

F. ff., 109, 110
Facsimile, 19
Fascicle (*Official Records*), 68 n.
Federal documents, laws, etc. See *under* U.S.
Figures (illustrations,) 113, 115
Folded materials, 130
Footnote citation, 1–4, 81–112; as evidence of quality, 81–82, 119; corporate author omitted in, 86, 90, 92, 95, 96, 108; differentiated from bibliographical references, 1–2, 81, 85; full forms (for first reference,) 84–105; full forms repeated, 106 and n.; modification of bibliographical forms for, 84–85; in science, 113–118;

Index

of source of indirect quotations, 88–89, 122; use of superior numerals, letters, and symbols, 82–83, 128, 129, 130 *and* 137–138, 143, 145–146
———— Alternatives to, 114–116. *See also* Terminal bibliography
———— Shortened forms (for second or later reference), 110–114, 111 (Bible,) 111–112 (classical and literary works), 2, 100 ff. (law), 2, 113 (science), with *ibid.*, *op. cit.*, and conventional abbreviations, 110 ff.
———— Uses of, 1–2, 81
Footnotes: arrangement of, 82–83, 84, 85, 127–129; in manuscripts for publication, 123, 129, 130 (double spacing required); in scientific literature, 114–116, 116–118 (examples in scientific journals); continuing on succeeding page, 128; location of, 83, 128 (bottom of page), 5, 110 n., 128–129 (end of chapter, etc.), 129 (scattered in text), numbering for, 82, 128; typing instructions for, 83, 127–130; use of symbols, 82–83. *See also* specimen pages, 137–138, 142–145, 147
Foreign government documents. *See* Government documents, foreign; Gt. Brit., government documents; Regional organizations; Specialized agencies
Foreign phrases, words, terms, 7, 111
Formulae (in text), 83
Front matter. See Preliminary material (pages)

Globe edition of Shakespeare's works, 12, 112
Glossaries, 130
Government documents: bibliographical references to, 30–35 (under U.S., but applicable to all documents); footnote citations to, 90–91, 105 ff.; references to foreign, 49–50. *See also under* Gt. Brit.; United Nations; U.S.; etc.
Government Printing Office. *See* U.S. Government Printing Office
Graphs (illustrations), 15, 82, 126
Gt. Brit. government documents: bibliographical references to, 43–49; Historical Manuscripts Commission publications, 46; laws, statutes, etc., 48; non-Parliamentary publications, 46–47; *Parliamentary Debates*, 47; Parliamentary publications, 43–46, 44 (Command Papers), *See also under* Court reports
———— Footnote citations to, 91–92; laws, etc., 92; *Parliamentary Debates*, 99–100
Gt. Brit., H.M. Stationery Office, 43, 46, 47
Greek authors. *See* Classics

Half title, 11 (series), 126
Hanging indention, 16, 23, 130, *and see* 139, 146, and "Selected List of References," 149–150
Hansard's Debates. See Parliamentary Debates
H.C. (House of Commons Papers), 43
Headings, contents and style of, in manuscripts, 122, 125, 127 *and* 136, 141–143
Hearings (U. S. Cong.), notation of, 34
H.L. (House of Lords Papers), 43
H.M. Stationery Office. *See* Gt. Brit., H.M. Stationery Office
Historical Manuscripts Commission. *See under* Gt. Brit.
"Home rule" charters, 42
Honesty, requirements of, in citation, 82. *See also* Acknowledgements
House of Commons (and/or

Lords) Papers. *See* Parliamentary publications *under* Gt. Brit., government documents

House of Representatives, bills, resolutions, reports, etc. *See* Congressional publications *under* U. S. government documents

Ibid., use of, 106–108, 110

Illustrations: in manuscripts and dissertations, 126; located in appendix, 130; notation of, in references, 15, 113 (science). *See also* Charts; Plates; etc., *and* 144

Illustrator, 10, 15

Imprint (place: publisher, date) in bibliographical references to: books, 12–14, 17 nn.; British documents, 45, 46; International Court of Justice, 74; International Labor Organization, 54; League of Nations, 52, 67 n.; omitted in legal literature, 41; Permanent Court of International Justice publications, 55; regional organizations, 79; specialized agencies, 76; state and municipal documents, 41; United Nations publications, 66; U.S. government documents, 32; use of n.d. and n.p. in, 13, 14

——— in footnote citations, 84–85, 86, 108; omission of, 111–112 (Biblical, classical, and literary references); 101 (legal references); 106 (shortened form for second or later use); used to identify corporate author, 86, 90, 92, 108

In, in his: use of, 28, 88

Indentions. *See* Hanging Indention; *also under* Typing the manuscript

Indexes, 125–126; preparation of, 130–131; typing instructions for, 131

Initials, author's, 7; following encyclopedia articles, 26; following preface, 125

Intergovernmental agencies. *See* Regional organizations; Specialized agencies

International agreements. *See* Treaty Series

International Court of Justice, 62, 73–74, 98

International Labor Organization, 50 (League of Nations), 75 (U.N.); bibliographical references to publications, 54–55, 76; footnote citations to publications, 98–99. *See also* Specialized agencies

International Law Commission (U.N.), 74

International Organizations. *See* League of Nations; Permanent Court of International Justice; United Nations; etc.

Interviews, 89

Introduction, alternative treatments of, 127

Italics, use of, for: abbreviations, foreign words, phrases, 111; Biblical, classical, and literary references, 111–112; document references, 31, 35, 37 (alternate form of document reference), 40, 68, 97, 99; legal citation, 76, 100–102; names of encyclopedic works, 26; names of periodicals, 20, 23, 87; scientific references, 114, 115; series notation, 11 (books), 37, 74, 94; titles of books, 9–10, 15–16, 85; titles of unpublished dissertations, 18

Italics in manuscripts: indicated by underlining, 9–10, 20, 115, 122; need for consistency, 119; for side headings, 127; title pages, 125; table of contents, 126. *See also* specimen pages, 135 ff.

Joint authors, citation of, 7

Journals. *See* Periodicals

Index

King James version of the Bible in footnote references, 112
Latin authors. See Classics
Law, citation in, 2, 103–104
Leaders, 126 and 136, 141, 142
League of Nations: official languages of, 50; provisions of Covenant of, 53–54, 55; transfer of functions to United Nations, 50, 57
——— Publications: bibliographical references to, 51–53; footnote citations to, 92–93; *Official Journal*, 51, 92–93; periodicals, 53, 93; sales numbers and symbols, 51–53. See also *Treaty Series*
Leaves (numbered), 110; designation of, in unpublished dissertations and manuscripts, 19, 89
Letters, citation of, 19–20, 89–90
Letters (superior), for footnote identification, 82–83, 114; in charts and tables, 126. See also specimen page 144
Library card catalog, 6, 9, 11, 30
"Literature Cited," 5, 115 and see 149–150.
Loc. cit., uses of, 106, 107–108, 110
L.S., 89–90

Magazines. See Periodicals
Manuscripts: bibliographcial references to, 19–20; footnote citations to, 89. See also Gt. Brit., Historical Manuscripts Commission
——— For publication: allowance for editorial comment, etc., 123; author's responsibility for all details, 119; citation should be in style of subject field, 3, 113, 118; corrections, 121; typing of footnotes, 129–130; importance of consistency of details, 119; index, 130–131; page numbering, 124; preface (function and contents of), 125; materials, 121; typing requirements, 123. See Italics, etc. *and* specimen pages 135 ff.
Maps, notation of, 15, 130 (appendix)
Margins, 121–122; for quotations, 122 and 137, 138
Mathematics, citation in, 83, 114, 119 and 147
Medicine, references in, 113
Microfilming, 120
Mimeographed documents. See Processed documents
MS, 20 n., 110
Municipal documents: bibliographical references to, 40–43 *passim*; footnote citations to, 90–91; ordinances and charters, 42–43
Musical scores, notation of, 15

N.d. (no date), 14, 110
Newspaper articles: bibliographical references to, 24–25; footnote citation to, 86, 87–88
Nobility, titles of. See Titles of nobility
Non-Parliamentary publications. See under Gt. Brit., government documents
N.p. (no place), 3, 110
N.s. (new series), 21, 110
Numerals (superior) for footnote identification, 82–83, 128, 129, 130 and 137–138, 143, 145

Official Journal (L. of N. and *Supplements* and *Annexes*). See under League of Nations
Official Records (U.N. and specialized agencies; *Annexes* and *Supplements*). See under United Nations
Omissions. See Ellipses
Op. cit., uses of, 106, 107–108, 110
Ordinances, city, 42–43
Out-of-print books, 13–14

Page numbering: in dissertations, 123–124; in publishers' manuscripts, 124; in reprint editions, 14; preliminary pages, 16 n., 123; *See also* Typing the manuscript

Pagination: In bibliographical references to: articles, 22–23, 25 (periodical), 24–25 (newspaper), 26–27 (encyclopedia); books, 14–15, 27–28 (parts of books or sets); *Congressional Record*, 37–38; government documents, 33; omitted in legal citation, 38; unpublished dissertations and manuscripts, 19

―――― Footnote citation of: articles, 86–87 (periodical), 88 (encyclopedias and parts of books), 87–88 (newspaper); Biblical, classical, etc., 111–112; books, 84, 85, 86; *Congressional Record*, 99–100; legal works, 101, 103 n., 105 nn.; scientific literature, 114 ff.; unpublished dissertations and manuscripts, 89. *See also* Preliminary pages

―――― In shortened, later forms of footnote citation, 106, 107

Paper: suitable quality of, for manuscripts, 121; marking of, to allow for footnotes, 128

Paragraph form (footnotes), 84, 85, 127, *and see* specimen pages 137 ff.

Parentheses, uses of, in citation: of imprint in footnotes, 84, 85; of newspaper articles, 24 (place), 25 (editorials); to enclose date, 26 (encyclopedia articles), 38, 39, 100 ff. (legal citation), 47 (*Parliamentary Debates*) 23, 24 (periodical articles), 38 (*Congressional Record*); to scientific works, 114 ff.

Parliamentary Debates: bibliographical references to, 47; footnote citation to, 100

Parliamentary publications. *See under* Gt. Brit., government documents

Periodicals: abbreviation of names, 87, 106, 114; change of names of, 21; names of, italicized in references, 21, 87; names of, not used with *op. cit., loc. cit.*, 108

―――― Articles in: bibliographical references to, 21–25; examples of references to, in science, 116–118; footnote citations to, 86–87. *See also* Articles *and under* U.S. Government documents, etc.

Periods: Biblical, classical, and literary references, 111 ff.; examples of use of, in bibliographical references, 16 (books), 23 (articles); in footnote citation, 85 (books), 87 (articles); not used in headings, 127; title pages, 125; used as leaders, 126 *and see* 136, cf. 141–142

Permanent Court of International Justice publications: bibliographical references to, 55–56; footnote citations to, 94. *See also* International Court of Justice

Personal author, citations of, in documents, 30–31, 36, 47 n., 66 (U.N.), 91, 102 n. (U.N.)

Ph.D. dissertation. *See* Dissertations

Photographs (illustrations), 126

Photostatic copies: for illustrations, 126–127; noted in manuscript references, 19

Pica typewriter type, 121, 123

Place of publication (in imprint), citation of: books, 12–13, 84, 85; government documents, 32 (U.S.), 41 (state), 45 (British), 54 (ILO), 57 (U.N.P.C.), 66 (U.N.), 76 (specialized agencies)

―――― Omission of, in references to: the Bible, classical literature, etc., 111; encyclopedic works, 26; legal works, 41, 101 ff.; peri-

Index

odical articles, 23, 87. *See also* Imprint

Plates (illustrations), notation of, 15 *and see* example, 16; numbering of, 126

Poetry, quotations from. *See* Quotations

Preface: date of, used in imprint, 14; functions and content of, 125; page numbering of, 123

Preliminary matter (pages), notation of, 15, numbering of pages in dissertation, 123; in manuscripts intended for publication, 124

Press releases (U.N.), 71

Printer's estimates, 123

Private Acts (U.S.), 39

Processed documents, 60 n., 60–61 (U.N.), 36–37 (U.S.)

Pseudonyms, 9

Public Acts (U.S.), 39, 101

Publisher's name: citation of, 12–14, 15–16, 85 (books), 32 (U.S. government publications); peculiarities of, 13

——— Omission of, in Biblical, classical, etc. references, 111–112 (cf. 4 n.); encyclopedia references, 26; legal citation, 39 ff., 100 ff.; periodical articles, 23, 85; state documents, 41. *See also* Imprint

Punctuation: modifications of, in footnote citations, 83; need for consistency of, 119; of author's name, 7; of dissertation and manuscript title pages, 124–125; of tables of contents, 126; of imprints, 12; of publisher's names, 13; of scientific references, 114 ff.; of subtitles, 10; of titles, 10; omission of, in page numbering, 124. *See also* Ellipses; Parentheses; Periods, etc.

——— In bibliographical references to: books, 15–16; periodical articles, 22–23

——— In footnote citations to: books, 84–85; Bible, 112; classical and literary works, 111–112; periodical articles, 86–87

Questionnaires, 130 (appendix)

Quotation marks, uses of: double and single, 122–123 *and* 137; omitted with single spacing, 122; with quotations of poetry in text, 123; to identify series, 11. *See also* Quotations

——— In titles in references to: articles, 23, 88 (encyclopedic works), 24–25 (newspapers), 20, 23–24, 87 (periodicals); government documents, 33, 39; statutes, 39 n., 101, 102; *Official Journal* (L. of N.), 92–93; *Official Records* (U.N., *Supplements and Annexes*), 68–69, 97; parts of books or sets, 27–28, 88; scientific works, 114; unpublished dissertations and manuscripts, 19–20

Quotations: acknowledgment of sources, 82, 122; arrangement of and indentions in manuscripts, 122–123, and 136, 137; direct, 14, 82, 122; from Biblical, classical, and literary works, 111–112; identification of omissions and interpolations in, 109; in the text, 72, 75, cf. 151; use of [*sic*], 111 *and* 136, 137; use of slash (virgule), 123

References, need for full notation of, 3–4, 113

Regional organizations, 77–80; bibliographical references to publications, 79

Registry number (treaties), 72, 98

Registration date (U.N. documents), 66 n.

Regnal year (English statutes), 48,

102
Reprint editions, 14
Roman numerals, uses of: for page numbering (lower-case), 16 n., 123; in sales numbers, 51 (L. of N.), 64 (U.N.); in tables of contents, 125; to differentiate page numbers, 84, 85; to number illustrations, 126; with abbreviations, 109–111

———— To indicate chapter numbers in manuscript, 125, 127; to indicate illustrations in manuscripts, 126; to indicate volume numbers of books in footnote citations, 85; encyclopedic works, 26; periodicals, 21, 87; *Treaty Series*, 93. *See also* specimen pages 136, 141, 142, 143

Sales numbers: International Court of Justice, 74; League of Nations Publications, 51–52; United Nations Publications 64–65, 66–67, 96, 96–97 n.
Science: citation in, 2, 83, 113–118; alternatives to footnote citation in, 114–116; biological science, 113; examples of, in journals, 116–118; importance of date in, 113; terminal bibliography used in, 119–122; use of symbols in, 82–83
Secretariat (U.N.). *See under* United Nations
"Selected List of References," 5, 149–150
Semicolon, 84 (imprint in footnotes), 10 (subtitles)
Senate bills, documents, resolutions, etc. *See* Congressional publications *under* U.S. Government documents
Serial numbers (U.S. documents), 32–33
Series, notation: in bibliographical references to books, 10–12; British documents, 44, 45; English court reports, 48; International Court of Justice, 73; International Labor Organization, 54; League of Nations, 53 n.; *Parliamentary Debates*, 47; periodical articles, 21, 23 n.; Permanent Court of International Justice, 55–56; specialized agencies, 76; state and municipal documents, 40; U.N. Preparatory Commission, 58; Secretariat (U.N.), 71; United Nations, 66 (doc. symbol replaces series); U.S. government documents, 11, 31, 33, 34, 36

———— In footnote citation: books, 84; British documents, 92 nn.; English court reports, 105 nn.; federal, state, and municipal documents, 90; International Court of Justice, 98; League of Nations, 92; *Parliamentary Debates*, 100; Permanent Court of International Justice, 94; specialized agencies (U.N.), 99; Conference agencies (U.N.), 99; U.N. Conference on International Organization, 95; United Nations, 96, 96 nn.

———— May replace corporate author in footnote citations, 86, 96
Sessional papers. *See* Parliamentary publications *under* Gt. Brit.
Shakespeare, citation of works of, 12, 112
Short Title Act (England), 102
Short Titles (International Court of Justice), 73 n.
Shortened forms of footnote citation. *See under* Footnote citation
Sic, 111 *and* 137, 138
Slash, 23 n., 61, 62, 67 n., 123 (poetry quotations)
Slip laws, 39 (U.S.), 48 (English)
Spacing. *See under* Typing the manuscript
Specialized agencies (U.N.), 75–77, 98–99; cf. (L. of N.); agree-

ments with U.N. negotiated, 58; cooperation coordinated by ESCO, 75; organization and functions, 75, 75 n., 76 n.
—— Publications, bibliographical, references to, 76, 77 (periodicals); footnote citation to, 98–99 (incl. periodicals); not to be confused with U.N. publications, 99
Specimen pages of dissertations (typewritten), bibliographies, 139, 146; chart, 144; list of charts, 142; tables of contents, 136, 141; title pages, 135, 140; text and footnotes, 137, 138, 143, 145, 147 (mathematics)
Spelling, consistency in, 119
Square brackets, uses of, 6, 19 *and* 145; in imprint, 12, 14; in scientific references, 114, 115
State bluebooks, 41
State documents: bibliographical references to, 40–41; footnote citations to, 90–91
State laws, statutes, etc.: bibliographical references to, 41–42; footnote citations to, 101–102, 104
Subject fields, need for conformity to citation practices in, 2–3, 83–84, 113, 118, 122–123
Subtitles: books, 10, 84, 105
Supplementary notation of illustrator, editor, etc., in references, 10, 84–85
Supplements, Official Records. See *Official Records* (U.N.)
Supreme Court reports. *See under* Court reports
Symbols, use of: document identification, 44 (Command Papers), 51–52 (League of Nations), 58, 59, 61–64 (U.N.), 94–95 (U.N.-C.I.O.), 58, 95 (U.N.P.C.); footnote citation, 82–83; references in charts, tables, etc., 126, *and* 144; scientific references,

82–83 *and* 147. See also 114–116; Sales numbers and "Symbols and Abbreviations," 109

Table of contents: functions and contents, and typing instructions for, 125–126; page numbering of, 123; use of leaders in, 126 *and* 136, 141, 142
Tables (illustrations): notation of, in citation, 15; superior letters in references to, 82; instructions for numbering and typing of, 126. *See also* specimen page 144
Technology, citation in. *See* Science, citations in
Terminal bibliography, 5, 106 n., 113–116, 118
Textbooks, the bibliography in, 5, 128
Theses. *See* Dissertations
Title, collective, 28
Title entry: for anonymous works, 9; articles, 24 (periodical), 25 (newspaper), 26 (encyclopedia); books, 18; *Congressional Record*, 38, 99; court reports, 48–49 (English), 42 (state), 39 (U.S.); *Parliamentary Debates*, 47; statutes, 38 (U.S.), 41 (state), 48 (English); *Treaty Series*, 53, 93, (L. of N.), 72–73 (U.N.); works of Shakespeare, etc., 111–112
—— Omission of corporate author in footnote citation with: books, 86; League of Nations publications, 92; United Nations publications, 96; U.S. government documents, 90–91
Title page: as source of bibliographical information, 6, 9–10, 11 (half-title), 12–14; contents and typing instructions for, 124–125; page numbering of, 123–124
—— Requirements of: for dissertations, 124–125; for publish-

ers' manuscripts, 125. *See also* 135, 140

Titles, in footnote citations: omissions from, indicated by ellipses, 31, 92, 109; repetition of, in a bibliography, 16 n.; replaced by *ibid., loc. cit., op. cit.,* 106–108; shortened, 103; meaning of, in law, 100, 101 nn.

—— Italicized in references to: books, 9–10, 15–16; collections of statutes (codes), law reports, etc., 38, 39, 41, 100–102; government documents (works complete in themselves), 31; names of encyclopedic works, 26; names of periodicals, 20; unpublished dissertations, 19–20

—— Placed with quotation marks in citations to: articles, 20, 22–23 (periodicals), 24–25 (newspapers), 26 (encyclopedia); government documents (parts of publications or periodicals), 31, 37; manuscripts, 19–20; parts of books or sets, 27–28; unpublished dissertations, 19. *See also* Alternate form of document reference

Titles (professional), omission of, 7, 124

Titles of nobility, 8–9

Translator, 8

Treaty Series: bibliography references to, 45 n. (Gt. Brit.), 53 (L. of N.), 72–73 (U.N.); footnote citations to, 92 n., (Gt. Brit.), 93 (L. of N.), 98 (U.N.); various *Series* should be identified, 53

Typing the manuscript: appendix materials, 130; bibliography, 130 *and* 139, 146; check for consistency and accuracy of all details, 119; corrections and erasures, 121; footnote styles, 83, 127–130 *and* 137, 138; cf. 114–116; indentions, 122–123; hanging indentions, 16, 23, 130 *and* 139, 146; headings, 127; index, 130, 131 *and* 151 ff.; margins, 121–122 (of quotations) 123; number of copies, 120; quotations, 122–123; page numbering, 123–124; proper materials, 121; spacing, 122–123; specimen pages of typewritten text, bibliography, etc., 135–147; table of contents, 125–126 *and* 136, 141; title page, 124–125 *and* 136, 140; illustrations, 126. *See also* Capitalization; Italics, use of

United Nations, 58–72, 95–97; Charter, 57, 58, 61, 70 (Secretariat), 75 (specialized agencies), 78 (regional organizations); complexity of, 59, 66; knowledge of structure, functions, etc., necessary, 59, 70 (Secretariat), 75 (specialized agencies); languages (official and working), 61; reference tools, 59, 59 n., 60 n., 75 n., 76 n.; transfer of League of Nations activities, 57. *See also* International Court of Justice; Regional organizations

—— Bibliographical references to publications, 65–67, 65 n.; document symbols, 66; imprint, 66; personal author, 66; periodicals, 74–75; registration date, 66 n.; sales number, 66–67, 67 n.

—— Documentation system (symbols), 58, 61–64; acronyms, 63; basic symbols (principal organs), 61–63; secondary symbols, 63–64; descriptive symbols (additive), 64; whole symbols, 63–64, 66; symbols of two organs, 67 n. *See also* International Court of Justice; U.N. Secretariat

—— Footnote citations to publications, 95–97, 99

—— *Official Records, Supple-*

ments, Annexes, 68–70, 97; bibliographical references to, 68; document symbol retained, 61; footnote citation, 97; *not* a sales publication, 64; *Supplements and Annexes,* 69–70, 72 n.
—— Press releases, 71
—— Sales publications and sales numbers, 64–65, 66–67, 96, 97; follows League of Nations practice, 64
—— Secretariat, 70–72; bibliographical references to publications, 72; documentation systems (peculiar to Secretariat), 71; functions, 70; treatises, 72
—— Specialized agencies, 58, 62, 75–77, 98–99; agreements with U.N., 58, 75; bibliographical references to publications, 76; footnote citation to publications, 98–99; periodicals, 77, 99; relation to U.N., 75
United Nations Conference on International Organization, 56–57, 94–95
United Nations Preparatory Commission, 57–58, 95
United Nations Treaty Series, 72–73, 98
U.S. government agencies, change of names of, 37
U.S. government documents, 29–39, 90–91, 99–100; government agencies/as corporate authors, 29–30; personal author, 30–31; serial numbers, 32–33. *See also* Codes (law); Court reports
—— Bibliographic references: Alternate form of reference, 37; bureau publications, 35; Congressional publications, 29–30, 32, 33–34; *Congressional Record,* 37–38; departmental publications, 35–37; laws, statutes, etc., 38–39; periodicals (published by U.S.), 37; personal author, 30–31
—— Footnote citations: various types of, 90–91; *Congressional Record,* 99–100; Constitution, 101; laws, statutes, etc., 100–101; personal author citation, 91
U.S. Government Printing Office, 43; notation of, in imprint, 13, 32
U.S. laws, statutes, etc., 38–39, 100–101. *See also under* Court reports
U.S. Supreme Court Reports. See under Court reports
University Microfilms, Inc., 120
Unpublished works. *See* Dissertations; Manuscripts

Verse, 111, 112, 123
Vide supra, 109, 111; cf. *infra,* 110
Virgule. *See* Slash
Volume: abbreviations, capitalization, and use of, 15, 23, 84, 111; omission of, in references, 84
Volume number, uses of: in bibliographical references to books, 14–15, *Congressional Record,* 38; encyclopedic works, 26; *Parliamentary Debates,* 47; Parliamentary publications, 44; periodical articles, 21, 23, 24; scientific literature (examples from journals), 116–118
—— In footnote citations of: books, 84–85, 86; *Congressional Record,* 99–100; encyclopedic works, 88; laws, statutes, etc., 100–102; *Parliamentary Debates,* 100; Parliamentary publications, 92 nn.; periodical articles, 87

World Court. *See* Permanent Court of International Justice

Year Books (Gt. Brit), 48, 104

www.ingramcontent.com/pod-product-compliance
Lightning Source LLC
Chambersburg PA
CBHW051614230426
43668CB00013B/2106